Contaminated Land and its Reclamatio

ISSUES IN ENVIRONMENTAL SCIENCE AND TECHNOLOGY

How to obtain future titles on publication

A subscription is available for this series. This will bring delivery of each new volume immediately upon publication. For further information, please write to:

The Royal Society of Chemistry
Turpin Distribution Services Limited
Blackhorse Road
Letchworth
Herts SG6 1HN, UK

Telephone: +44 (0) 1462 672555
Fax: +44 (0) 1462 480947

ISSUES IN ENVIRONMENTAL SCIENCE
AND TECHNOLOGY

EDITORS: R. E. HESTER AND R. M. HARRISON

7

Contaminated Land and its Reclamation

THE ROYAL
SOCIETY OF
CHEMISTRY
Information
Services

ISBN 0-85404-230-X
ISSN 1350-7583

A catalogue record for this book is available from the British Library

Published by The Royal Society of Chemistry, Thomas Graham House, Science Park, Milton Road, Cambridge CB4 4WF, UK

Typeset in Great Britain by Vision Typesetting, Manchester
Printed and bound in Great Britain by Bookcraft (Bath) Ltd

Preface

As with previous volumes in this series, we have drawn on advice from members of our distinguished Editorial Advisory Board to select a topic which is of particular current interest and importance, and a set of authors who are recognized authorities in the area. **Contaminated Land and its Reclamation** is an issue which has been the focus of much attention during the past decade in the UK, in continental Europe, in the USA, and more widely throughout the world. There have been many incidents reported in the popular media which have served to raise the level of public concern about the effects and consequences of land contamination and governments have moved to introduce legislation designed to control and regulate activities which lead to contamination and to deal with the historical legacy of contaminated sites.

This volume begins with an overview of the subject by Peter Young and Phillip Crowcroft of the well-known UK firm of consultants, Aspinwall & Company Ltd., and Simon Pollard of the Scottish Environmental Protection Agency. They deal with recent developments of a legal, technical, and professional nature and relate these to the methods of risk assessment and their application. The role of professional environmental consultants in providing specialist advice on site evaluation and remediation techniques is reviewed. Mary Harris of Monitor Environmental Consultants Ltd. and Judith Denner, head of the new UK Environment Agency's contaminated land function, have then set out in considerable detail the main features of current procedures in their article on UK Government policy and controls. They review the key developments in the UK over the past twenty years, including the Environmental Protection Act (EPA) 1990, the European Commission Green Paper on Environmental Liability, the consultation paper 'Paying for Our Past', the policy paper 'Framework for Contaminated Land', and the Environment Act 1995 and the associated draft statutory guidance (not yet released at the time of writing). Section 143 of the EPA 1990 imposed a duty on local authorities to compile registers of contaminated land throughout the UK and led to serious concern being expressed by property owners and others on the grounds of blight and negative effect on property values. The need to reach a balance between the wish to maintain and improve environmental quality on the one hand and to ensure that business and land owners are able to cope with the environmental costs on the other has informed subsequent legislation. The Harris and Denner article provides a fascinating insight into the evolutionary processes involved in determining policy and establishing controls in this area.

The article by Peter Wood of AEA Technology's National Environmental Technology Centre is concerned with remediation methods designed to clean up contaminated land. The treatment options reviewed here are landfill, containment, physical, chemical, biological, and thermal processes, and stabilization/solidification. Consideration of how these options can be combined in an integrated process is included, together with a discussion on selecting the best practicable environmental option. The particular example of land reclamation after contamination by coal-mining operations is examined by David Rimmer and Alan Younger of the University of Newcastle upon Tyne. The problems associated with both colliery spoil from deep mines and the disturbance associated with opencast or strip mining are outlined and techniques developed to overcome them are described. The most chemically focused and explicit article here is concerned with lead, zinc, and cadmium pollution and remediation methods for soils contaminated with these non-ferrous heavy metals. Jerry Schnoor and his colleagues from the University of Iowa and Kansas State University detail procedures for the *in situ* sequestration of these metals in insoluble phosphate minerals. The metals are common contaminants in superfund sites in the USA and phosphate remediation is shown to be particularly effective for lead.

The role of a regional development agency in improving the environment and the local economy through the reclamation of derelict and contaminated land is described by Gwyn Griffiths and Steve Smith of the Welsh Development Agency (WDA). The tragic loss of life which resulted in 1966 from a colliery spoil tip slide at Aberfan stimulated a major programme of derelict land reclamation in Wales, around 50% of this being unrelated to coal mining. High priority is given to sites having development potential and a robust system of management has been established which is set out in the WDA Manual on the Remediation of Contaminated Land. The guidance embodied in this document has wide-ranging applicability and is described here in some detail. Finally, the legal liabilities and insurance aspects of contaminated land are examined by Tony Lennon of ECS Underwriting, London. The financial risks associated with land remediation and reclamation processes are considerable and the standard form of public liability insurance commonly contains exclusion clauses for pollution-related damages (*e.g.* effects of asbestos in buildings, or methane seepage from landfill). However, the development of specific environmental policies has been hindered by technical problems; these led to the collapse of this sector of the insurance industry in the USA in the early 1980s and continue to impact on the UK industry even today. The way in which such problems have been overcome is discussed; this will be of great interest to land owners and developers and to those engaged in consultancy and contracting work related to the clean-up of contaminated sites.

Taken together, this set of seven reviews represents a broadly based, timely and authoritative treatment of the main issues concerning contaminated land and its reclamation. As such it is expected to make an important contribution to the public debate on these issues and to be essential reading for all those groups of people directly or indirectly involved, from consultants and their technical advisors, through developers, contractors, and land owners, to local authorities and government agencies with responsibility for policy and its implementation in this area.

Ronald E. Hester
Roy M. Harrison

Contents

Contents

Contents

Editors

Ronald E. Hester, BSc, DSc(London), PhD(Cornell), FRSC, CChem

Ronald E. Hester is Professor of Chemistry in the University of York. He was for short periods a research fellow in Cambridge and an assistant professor at Cornell before being appointed to a lectureship in chemistry in York in 1965. He has been a full professor in York since 1983. His more than 300 publications are mainly in the area of vibrational spectroscopy, latterly focusing on time-resolved studies of photoreaction intermediates and on biomolecular systems in solution. He is active in environmental chemistry and is a founder member and former chairman of the Environment Group of The Royal Society of Chemistry and editor of 'Industry and the Environment in Perspective' (RSC, 1983) and 'Understanding Our Environment' (RSC, 1986). As a member of the Council of the UK Science and Engineering Research Council and several of its sub-committees, panels and boards, he has been heavily involved in national science policy and administration. He was, from 1991–93, a member of the UK Department of the Environment Advisory Committee on Hazardous Substances and is currently a member of the Publications and Information Board of The Royal Society of Chemistry.

Roy M. Harrison, BSc, PhD, DSc (Birmingham), FRSC, CChem, FRMetS, FRSH

Roy M. Harrison is Queen Elizabeth II Birmingham Centenary Professor of Environmental Health in the University of Birmingham. He was previously Lecturer in Environmental Sciences at the University of Lancaster and Reader and Director of the Institute of Aerosol Science at the University of Essex. His more than 250 publications are mainly in the field of environmental chemistry, although his current work includes studies of human health impacts of atmospheric pollutants as well as research into the chemistry of pollution phenomena. He is a former member and past Chairman of the Environment Group of The Royal Society of Chemistry for whom he has edited 'Pollution: Causes, Effects and Control' (RSC, 1983; Third Edition, 1996) and 'Understanding our Environment: An Introduction to Environmental Chemistry and Pollution' (RSC, Second Edition, 1992). He has a close interest in scientific and policy aspects of air pollution, currently being Chairman of the Department of Environment Quality of Urban Air Review Group as well as a member of the DoE Expert Panel on Air Quality Standards and Photochemical Oxidants Review Group and the Department of Health Committee on the Medical Effects of Air Pollutants.

Contributors

P. Crowcroft, *Aspinwall & Company Ltd, Walford Manor, Baschurch, Shrewsbury, Shropshire SY4 2HH, UK*

J. Denner, *Department of the Environment, Romney House, 43 Marsham Street, London SW1 3PY, UK*

L. E. Erickson, *Department of Chemical Engineering, Kansas State University, Durland Hall, Manhattan, Kansas 66506, USA*

G. Griffiths, *Welsh Development Agency, QED Centre, Treforest Industrial State, Mid Glamorgan CF37 5YR, UK*

M. Harris, *Monitor Environmental Consultants Ltd, Blakeland House, 400 Aldridge Road, Great Barr, Birmingham B44 8BH, UK*

M. Lambert, *Department of Geology, Kansas State University, Manhattan, Kansas 66506, USA*

A. J. Lennon, *ECS Underwriting, IVEX House, 42–47 Minories, London EC3N 1DY, UK*

G. Pierzynski, *Department of Agronomy, Kansas State University, Throgmorton Hall, Manhattan, Kansas 66506, USA*

S. Pollard, *Scottish Environment Protection Agency, Erskine Court, The Castle Business Park, Stirling FK9 4TR, UK*

D. L. Rimmer, *Department of Agriculture, The University, Newcastle upon Tyne NE1 7RU, UK*

J. L. Schnoor, *Department of Civil and Environmental Engineering, University of Iowa, 1134 University Building, Iowa City, Iowa 52242, USA*

S. Smith, *Welsh Development Agency, Pearl House, Greyfriars Road, Cardiff CF1 3XX, UK*

Contributors

P. A. Wood, *AEA Technology, National Environmental Technology Centre, Culham, Abingdon, Oxfordshire OX14 3DB, UK*

P. J. Young, *Aspinwall & Company Ltd, Walford Manor, Baschurch, Shrewsbury, Shropshire SY4 2HH, UK*

A. Younger, *Department of Agriculture, The University, Newcastle upon Tyne NE1 7RU, UK*

Overview: Context, Calculating Risk and Using Consultants

PETER J. YOUNG, SIMON POLLARD AND
PHILLIP CROWCROFT

1 Introduction

Historical Perspectives

The presence of land contamination is an inevitable legacy of an industrial past. At least since the time of Roman lead mining, the UK has seen human activities leave naturally present contamination concentrated, or altered in its chemical form. Since the industrial revolution of the late 18th century, an increasing range of new compounds have been manufactured and added to the list of contaminants, and the affected area of land has increased dramatically. Current estimates suggest between 100 000 and 220 000 ha of land is contaminated,[1] representing between 0.4 and 0.8% of the total UK land area. Table 1 sets out the principal types of contamination and its typical severity for major industrial land uses, based on results from more than 500 site-based studies.[2]

Much uncertainty over the presence of contaminated land is derived from the lack of a consistent definition. Other designations of land such as vacant, derelict, industrial or damaged can be confused with evidence for contamination. Contaminated land is not the same as derelict land; nor is the term contaminated land applied in the strict scientific sense of indicating the presence of an introduced substance which is harmful. If the substance is harmful, it is a 'pollutant'.

The Department of the Environment (DoE) has described contaminated land as 'land which represents an actual or potential hazard to health or the environment as a result of current or previous use'. Contrast this with derelict land which is 'land so damaged by past activities that it is incapable of beneficial use without treatment'. Clearly the two are independent. A worked-out limestone quarry may be derelict but will not be contaminated. In contrast, many established chemical factories will be contaminated but not derelict. Contaminated land poses a potential threat to the environment, but derelict land is a threat to future development, although it may be aesthetically unattractive as well.

Against this complex background of definition, the public, and financial markets, became alerted to land contamination by a sequence of well publicized

---

[1] Parliamentary Office of Science and Technology (POST), *Contaminated Land*, POST, London, 1993.
[2] P. J. Young in *Construction Law and the Environment*, ed. J. Uff, H. Garthwaite and J. Barber, Wiley Chancery, London, 1994, pp. 213–235.

Table 1 Categorization of major industrial land uses*†

Category 1: high contamination	Principal type of contamination‡
Hazardous waste treatment	O/I
Bulk organic chemicals manufacture	O
Bulk inorganic chemicals manufacture	I
Fine chemicals manufacture	O
Coal gasification/carbonization	O/I
Landfill and other waste treatment/disposal	O/I
Steelworks	O/I
Lead metal ore processing and refining	I
Oil refining and petrochemicals production	O/I
Pesticides manufacture	O
Asbestos & asbestos products manufacture	I
Scrap yards	O/I
Pharmaceuticals manufacture	O

Category 2: moderate contamination	
Drum and tank cleaning/recycling	O
Fertilizer manufacture	I
Non-ferrous metal ore mining	I
Wood preservatives production & timber treatment	O
Docks	I
Electric/electrical equipment manufacture	O
Mechanical engineering	O/I
Garages/filling stations	O
Minerals processing (bricks, cement, tarmac, *etc.*)	I
Power stations	I
Sewage works/farms	I
Shipbuilding/shipbreaking	O/I
Textile production and dying	O
Tyre manufacture and other rubber processing	O/I
Metal (other than iron or lead) processing/refining	I
Pulp and paper manufacture	O
Paint and ink manufacture	O/I
Electroplating and other metal finishing	O/I
Precious metals recovery	I
Foundries	I
Tanneries	O/I

Table 1 Continued	*Category 3:* *slight contamination*	*Principal type of* *contamination*‡
	Timber products manufacture	O
	Animal processing works	O
	Glass manufacture	I
	Road haulage yards	O
	Building trades products manufacture	I
	Printing works	O
	Research laboratories	O
	Airports & airfields	O
	Vehicle manufacture	O
	Railway yards/sidings	O
	Toiletries, detergents, disinfectants, *etc.*, manufacture	O
	Electricity sub-stations	O
	Dry cleaners	O

*This categorization is for illustrative purposes. It will give a broad indication only of whether the business concerned involves a contaminative use and, if so, the degree of seriousness of that contaminative use, and nature of the contamination arising from a contaminative use.

†Whether a business falls within a particular category will depend on a number of further factors, such as: whether the previous business use of the site has given rise to contamination; the length of time for which the site has been used for the business purpose; the overall sensitivity of the site with respect to its broader environmental setting; assessment of the extent to which the business follows good environmental practices and management controls; the extent of the manufacturing or processing activity which is carried on by the business and the site; assessment of the influence of the underlying geology and its hydrogeological characteristics.

‡'O' signifies organic contamination; 'I' signifies inorganic contamination.

incidents in the 1980s and early 1990s (Figure 1). At this time, investigation for soil and groundwater contamination was merely costed for in the purchase of land for redevelopment, and prior investigation was frequently patchy or non-existent. Three events which were influential in focusing concern about contaminated land are summarized in Table 2.[1,3,4]

These and other sites were described to the House of Commons Environment Committee in 1989 and their report, published in 1990,[5] called for registers of contaminated land to be established. This greatly influenced DoE thinking on the Environmental Protection Act, which was at the final drafting stage. The DoE had previously aimed to provide guidance to developers through advisory limits for concentrations of contaminants through publications from the Interdepartmental Committee on the Redevelopment of Contaminated Land (ICRCL). The idea of registering known or potentially contaminated land was more interventionist

[3] Derbyshire County Council (DCC), *Report of the Non-statutory Public Inquiry into the Gas Explosion at Loscoe, Derbyshire 24 March 1986*, DCC, Derby, 1988, vols. 1 and 2.

[4] M. Sury and A. Slingsby, *Proceedings of the 2nd International Conference on Construction on Polluted and Marginal Land*, Brunel University, London, 1992, pp. 379–384.

[5] House of Commons Environment Committee, *Contaminated Land*, HMSO, London, 1990, vols. I, II and III.

Figure 1 Bungalow
demolished by landfill gas
explosion.
(Courtesy of the
Derbyshire Constabulary)

Figure 1 Bungalow demolished by landfill gas explosion. (Courtesy of the Derbyshire Constabulary)

and, if not handled extremely carefully, could blight urban areas where redevelopment and investment is most needed.

In 1990, the DoE published the Government's response to the Environment Committee.[6] Linked to this were three actions which injected momentum into the field of contaminated land assessment:

1. A provision (section 143) was made in the Environment Protection Act 1990 for regulations requiring local authorities to compile registers of potentially contaminated land.
2. The DoE initiated a process of consultation on registers and contaminated land policy.
3. The DoE began to expand its research programme to provide broader guidance on contaminated land.

Once a proposal for registers of potentially contaminated land was incorporated in legislation, a lengthy and active debate was triggered. The Government sought to focus and inform this debate by a series of consultation papers as new policy was developed. The DoE also informed itself through international liaison where alternative approaches have been offered, many of these coming under national criticism over expense or quality of environmental protection. An interpretation of the key points emerging during this debate is offered in Table 3. In understanding the current legislative position on contaminated land it is important to have regard to this recent period of debate and uncertainty, with the emergence of a true risk-based approach founded on firm scientific and

[6] Department of the Environment, *The Government's Response to the First Report from the House of Commons Select Committee on the Environment on Contaminated Land*, HMSO, London, 1990.

Table 2 Example
incidents at contaminated
land sites

Site	Date of incident	Description of problem	Costs/implications
Clarke Avenue, Loscoe, Derbyshire[3]	March 1986	Explosion due to landfill gas migration demolished bungalow, badly injuring the occupants	£325–375k;[1] continued difficulties with mortgage lending and insurance
Ilford, Essex[4]	Late 1980s	Low level radioactivity found on former chemical site being redeveloped for housing	£10 million approx.; extensive remediation and delays to site development
Lumsden Road, Portsmouth, Hampshire[1]	1991	120 families rehoused after Ministry of Defence development of 1950s/1960s constructed on former Royal Navy waste site found contaminated with asbestos and heavy metals	£6.12 million,[1] including rehousing, medical screening, demolition, remediation and rebuilding

toxicological principles. Further detail on the current regulatory framework and its foundation on risk assessment is given in Section 3.

Current Consequences

The current position with contaminated land is largely attributable to the response from Government, and more widely to the concerns highlighted at the beginning of this decade. These have led to a radical and rapid change in the technical approach to evaluating and remediating contaminated land. The routine introduction of risk assessment, a phased and chemically orientated approach to site investigation, and the selection of remediation strategies from more technologies than simple cap and cover systems are all rapidly developing areas described below in Sections 3–5. Underpinning these developments have been the changes in the legislative position, the growth in research based guidance, particularly from DoE, and the availability of specialist expertise from environmental consultants (see Section 6).

From a technical viewpoint, recent developments have been extremely beneficial. The complexity of evaluating contamination has been thoroughly exposed, and the risk associated with facile and inadequate approaches prevailing in the market place has been much reduced. The prospect of new guidance values for key contaminants which take account of the availability, toxicity, pathway and impact of the contaminant will be a significant improvement over existing ICRCL guidance. However, the draft guideline values only address human health. Separate and additional assessment techniques are required for ecological, phytotoxic and physical damage potential. The link has also been

Table 3 Key issues arising from UK contaminated land policy development since 1990

Date	Action	Description	Consequence
May 1991	Consultation Paper on Public Registers[7]	Proposals to introduce local authority administered registers of potentially contaminated land: 'land subject to contaminating uses'; 40 such uses identified	Initiated wide debate on use of registers and contaminated land in general; concerns over blight and application of 'polluter pays' principle expressed
July 1992	Revised Proposals for Public Registers	List of contaminating uses reduced to eight, representing perhaps 10–15% of area of 1991 proposals[1]	Concerns over land values and blight shown to be fundamental to principle of registers, not to their scope. Objectives of registers received wide support; their consequence did not
March 1993	Registers withdrawn; Government Policy Review initiated	An alternative to registers was sought to achieve improvement in knowledge and planning for contaminated land redevelopment and remediation	Many relevant professions now actively engaged in finding an acceptable policy solution
March 1994	*Paying for our Past*, Consultation Paper issued	Included a focus on the dilemma of the 'polluter pays' principle *versus* the large historical legacy of contaminated land	Enhanced recognition that a risk-based approach was needed to prioritize the few sites which caused actual environmental harm
November 1994	*Framework for Contaminated Land* published[8]	Reaffirmed UK 'suitable for use' approach which requires remedial action only where contamination poses unacceptable risks to health or the environment, taking into account land use and setting	Identified need to build on existing statutory nuisance powers as primary basis for dealing with contaminated land

Table 3 Continued

Date	Action	Description	Consequence
1995	*A Guide to Risk Assessment and Risk Management for Environmental Protection* published[9]	Risk Assessment promoted as key method of approach and later in the year new draft guideline values for the human health assessment of contaminants in land are developed based on toxicological and exposure pathway risk assessments	Risk-based approaches promoted by wide range of professions and development authorities start to become more consistent and authoritative
1995	s57 of Environment Act inserts new Part IIA into Environmental Protection Act on Contaminated Land	The new legislation aims to clarify which land should be regarded as contaminated, who should pay for necessary remediation, and remove blight through suspicion of future regulatory control	The new provisions are complex, requiring support from extensive statutory guidance
September 1996	Draft Statutory Guidance on Contaminated Land published for consultation[10]	The guidance provides specific definitions under the terms of the 1995 Act, and addresses the inspection (by local authorities), definition and remediation of contaminated land as well as apportionment of liabilities for, and recovery of, the costs of remediation	Continued active consultation (until December 1996) and highlighting of link to extensive technical guidance published by the DoE and others to support the identification, classification and remediation of contaminated sites

[7] Department of the Environment (DoE) and Welsh Office, *Public Registers of Land which may be Contaminated: A Consultation Paper*, DoE, London, 1991.

[8] Department of the Environment (DoE), *Framework for Contaminated Land: Outcome of the Government's Policy, and Review, and Conclusions from the Consultation Paper 'Paying for our Past'*, HMSO, London, 1994.

[9] Department of the Environment (DoE), *A Guide to Risk Assessment and Risk Management for Environmental Protection*, HMSO, London, 1995.

[10] Department of the Environment (DoE), Welsh Office and Scottish Office, *Consultation on Draft Statutory Guidance on Contaminated Land*, DoE, London, 1996, vols. 1 and 2.

made between the soil and water environments,[11] so that contaminated land assessment and remediation should succeed in addressing all potential impacts in one overall risk-based approach. Such an integrated approach places new demands on the quality of the advice available on contaminated land. Awareness of the rapidly developing technological base is required, whether in investigation, risk assessment or remediation. A common understanding and language must be used to transfer knowledge clearly between chemists, engineers, environmental scientists, chartered surveyors, developers and lawyers, to name a few of those whose skills are needed to deal successfully with contaminated land. It is against this background of escalating demand that this article has been produced.

2 Identifying Contamination

To contaminate is defined as 'to make impure or pollute' in the Oxford English Dictionary. For the purposes of dealing with land, contamination is now taking on a dual meaning. The first is defined by legislation in the Environment Act 1995, and relates to the potential to cause significant harm to humans or pollution of controlled waters. In a wider sense, however, contamination may be viewed as a condition whereby soil or water contains above background concentrations of substances which are not normally there. These can be both chemical and physical, but for the purposes of this review, chemical contamination is most relevant.

Defining soil as contaminated is not merely a question of whether a chemical substance, say cadmium or phenol, is present, but whether it is present in sufficient quantity to cause harm. In relation to the Environment Act 1995, harm is defined, *inter alia*, as death, serious injury, disease, genetic mutation or birth defects. A great deal of cadmium or phenols are needed in soil to cause death and, furthermore, death will only occur if the soil is ingested in large quantities. So this raises the question of whether the mere presence of a chemical in soil at high concentrations is contamination. The answer in the UK is that such soil is contaminated, but the land would only be defined as 'contaminated land' under the regulatory regime if harm was occurring or likely to occur.

Clearly then, contamination of soil or water may exist, but if it is not causing harm it is possible to consider leaving it alone until such time as a particular site needs to be developed or reclaimed.

Contamination can create a range of hazards, depending on its composition and nature. It may be present in solid, liquid or gas phases, and may be physical, chemical or biological. Contaminants that are hazardous to humans are not necessarily hazardous to building materials or flora. Hazards may include:

- carcinogenicity
- toxicity
- asphyxiation
- corrosivity
- phytotoxicity

[11] Department of the Environment (DoE), *A Framework for Assessing the Impact of Contaminated Land on Groundwater and Surface Water*, CLR1, DoE, London, 1994, vols. 1 and 2.

- combustibility
- explosivity
- radioactivity

The process of assessing whether the existence of contamination matters is termed risk assessment. This is discussed more fully in the next section. Of prime importance in carrying out a risk assessment is the need to identify what or who is at risk, and these risk groups are termed targets or receptors. They comprise the following:

(i) People
- adults
- children
- site workers (temporary and permanent)
- permanent residents
- visitors
- neighbours

(ii) Water resources
- surface water
- groundwater

(iii) Flora and fauna
- sites of special scientific interest (SSSIs)
- livestock and wild animals, birds, *etc.*
- landscaping

(iv) Buildings
- foundations
- services
- structures

It should be remembered that not all the above receptors will exist at all locations, and part of the risk assessment process will be to define site-specific receptors.

3 Risk Assessment

Principles

With introduction of the new regulatory regime for contaminated land in section 57 of the Environment Act 1995, the UK has endorsed a risk-based framework within which contaminated land can be identified, assessed and managed.[8] This framework adopts a rationale and terminology from accepted international approaches to the regulation of environmental media and it embodies the fundamental distinctions between toxicity, hazard and risk that exist throughout environmental science. These terms are defined below:

- toxicity: the potential of a material to produce injury in biological systems
- hazard: the nature of the adverse effect posed by the toxic material
- risk: the probability of suffering harm or loss under specific circumstances

The term 'risk' has a multitude of uses and is not to be confused with 'hazard'. In the context of contamined land, the term 'risk' is used widely across disciplines when referring to issues such as first and third party financial liability, risks to human health and the environment, the perceived consequences of chemical exposure and the operational risk of project over-run. There is always a requirement, therefore, to state clearly what form of risk is under consideration and what its components are.

In terms of contaminated land, risks to human health and the environment can be regarded as being comprised of the following components:

- a source: a toxic substance or group of toxic substances with the potential to cause harm
- a pathway: a route by which a receptor could be exposed to, or affected by, the toxic substance(s)
- a receptor: a particular entity which is being harmed or adversely affected by the toxic substance(s)

Hazards arising from chemical exposure are characterized specifically by the nature of the adverse effect, the receptor and the target they affect; like physical hazards, they can only be realized where there is a 'linkage' between the source, the pathway and the receptor. The probability of a hazard being realized, *i.e.* the risk, depends on the context of this linkage, including site-specific factors such as the contaminant concentration in the exposure medium, its bioavailability, the ease of access to the exposure pathway and the duration of exposure. The consequences of the risk under consideration depend on site-specific factors such as the toxicological potency of the contaminant being considered, the specific adverse effect on the receptor, the duration of exposure and the sensitivity of the receptor (*e.g.* child *versus* adult, sites of special scientific interest (SSSIs) *versus* unprotected ecosystems of low sensitivity).

Risk Assessment. The process of risk assessment can be defined as simply 'an evaluation of the probability of harm'[12] and, in the context of contaminated land, is concerned with gathering and interpreting information on the characteristics of sources, pathways and receptors at a specific site and understanding the uncertainties inherent to the ensuing assessment of risk. The requirements of the risk assessment set the scope of a site investigation and, together, these activities form the scientific part of the contaminated land investigation. In practice, this involves characterization of the environmental chemistry of the contaminants, relevant properties of the soil(s) encountered and the wider site characteristics that influence contaminant fate and transport.

The environmental properties of chemicals encountered at contaminated land sites (Table 4)[13] determine the distribution, fate and transport of the contamination and play a major role in determining:

---

[12] L. D. Hooper, F. W. Oehme and G. R. Krieger, in *Hazardous Materials Toxicology*, ed. J. B. Sullivan, Jr., and G. R. Kreiger, Williams and Wilkins, Baltimore, 1992, pp. 65–76.
[13] S. J. T. Pollard, R. E. Hoffmann and S. E. Hrudey, *Can. J. Civil Eng.*, 1993, **20**, 787.

Table 4 Environmental fate and transport properties of selected contaminants

Compound	Aqueous solubility (mg l^{-1})	Vapour pressure (Pa)	K_{oc}*	Dominant partition medium
Benzene	1780	1.3×10^4	65	Air
Phenol	8.2×10^4	71	14	Water
Hexachlorodibenzo-*p*-dioxin	1.3×10^{-4}	1.9×10^{-6}	2.6×10^7	Soil
Benzo[*a*]pyrene	3.8×10^{-3}	7.3×10^{-7}	5.5×10^6	Soil

*Organic carbon–water partition coefficient.

(i) the suite of chemical analyses undertaken as part of the site investigation
(ii) the exposure assessment component of the risk assessment
(iii) the screening of remedial technologies

Because soil is itself a multi-media environment comprising solid, liquid, gaseous and biotic components, understanding the relative distribution and flux of contaminants between these components is an essential prerequisite to the site investigation, risk assessment and remedial plan. For example, the 4–6 ring polynuclear aromatic hydrocarbons (PAHs) possess very high soil organic carbon–water partition coefficients and remain predominantly adsorbed to the solid soil matrix. Therefore, where high molecular weight PAHs are of concern, analytical efforts focus on soil-bound PAH; the exposure assessment centres on pathways that involve the ingestion and inhalation of soil particles; and the remedial plan considers specific technologies that treat the soil matrix, such as soil-phase bioremediation.

The process of risk assessment[14] can be viewed as consisting of four key stages:

- hazard identification (what are the hazards presented by toxic substances at the site?)
- exposure assessment (what are the key environmental pathways and exposure routes by which the toxic substances can reach the receptors and, if required, what are the concentrations of substances at the point of exposure?)
- dose–response assessment (how potent are the toxic substances that can reach the receptor?)
- risk characterization (what level of risk can be assigned to each source–pathway–receptor linkage?)

Methodologies for contaminated land risk assessment vary widely between practitioners and jurisdictions but can essentially be categorized as qualitative, semi-quantitative and quantitative in their approach. In most cases, a qualitative assessment of risk is sufficient to identify the key issues at a contaminated site, providing it includes the full range of toxic contaminants encountered, takes account of direct and indirect exposure pathways and considers relevant receptors on and off the site.

[14] S. J. T. Pollard, D. O. Harrop, P. Crowcroft, S. H. Mallett, S. R. Jeffries and P. J. Young, *J. Chart. Inst. Water Environ. Manage.*, 1995, **9**, 621.

Where the source–pathway–receptor linkage is established, the qualitative approach can usefully provide an initial ranking of risks as insignificant, low, medium and high, depending on the site-specific factors mentioned above. These obviously are subjective designations and would require their own criteria in order to achieve uniformity of approach between sites (Table 5). Within the context of a tiered approach to risk assessment, semi-quantitative and quantitative risk assessment (QRA) methodologies are reserved for situations where greater resolution is required between risks in order to select between risk management options. QRA has become a highly specialized tool that relies heavily on the expert understanding and interpretation of baseline toxicological data. It can be applied only where contaminated sites are very well characterized and hazards are well defined (*e.g.* sites with radioactivity).

Risk Management. Risk management involves 'the evaluation of alternative options taking into account available economic, regulatory, political, social, scientific and technological information in order to select the most appropriate means of reducing risk' (adapted from La Grega *et al.*[15]). In practice, risks are managed by isolating or removing contaminant sources, intercepting exposure pathways or isolating or removing receptors; the guiding principle is to break the source–pathway–receptor linkage. The process of risk management employs the scientific output of the risk assessment, but considers other factors such as the financial and technical feasibility of remedial technologies, planning constraints and risk perception issues.

Environmental Criteria Development

Two recurrent themes with respect to contaminated land over the last 20 years have been 'when is a site contaminated?' and 'how clean is clean?'. Regulatory approaches to addressing these questions have, historically, relied on the development of environmental assessment criteria for soils and groundwaters, against which analytically determined site concentrations can be assessed. It is usual to take into account analytical detection limits and background concentrations in establishing such criteria. Three approaches have been dominant:

(i) the 'trigger' and 'action' guideline concentrations such as those developed by the UK Inter-departmental Committee on the Redevelopment of Contaminated Land (ICRCL), which considers two tiers of intervention: the 'trigger' value indicating when further investigation is necessary, and the 'action' value indicating where some form of remedial action is likely for the proposed end use of the site

(ii) the 'ABC' guideline values historically adopted by the Dutch Government and Ontario Ministry of Environment which represented background (A), intermediate (B) and heavy (C) levels of contamination for soil and groundwater

[15] M. D. LaGrega, P. L. Buckingham, J. C. Evans and the Environmental Resources Management Group, in *Hazardous Waste Management*, McGraw-Hill, Maidenhead, 1994, pp. 837–882.

Table 5 Example of a
site-specific qualitative risk
assessment matrix

Source	Hazard	Pathway	Receptor	Site-specific risk
Chromium(III)	Non-carcinogenic toxicity	Incidental ingestion	Humans, site fauna	Low to medium
		Inhalation	Humans, site fauna	Low to medium
		Dermal contact	Humans, site fauna	Low to medium
		Leaching	Watercourse	High
		Leaching	On-site buildings	Low
Chromium(VI)	Non-carcinogenic toxicity	Incidental ingestion	Humans, site fauna	Medium
		Inhalation	Humans, site fauna	Medium
		Dermal contact	Humans, site fauna	Medium
		Leaching	Watercourse	High
		Leaching	On-site buildings	Low to medium
	Carcinogenic toxicity	Inhalation	Humans, site fauna	Medium
Zinc	Phytotoxicity	Root uptake	Site flora	Low to medium
Suspended solids	Increased turbidity	Direct run-off	Watercourse	High
Methane	Explosion	Permeable strata	Buildings	High

(iii) the health risk assessment approach to site remediation enforced by the United States Environment Protection Agency (US EPA) under the 'Superfund' legislation which sets individual criteria, as 'applicable, relevant and appropriate requirements (ARAR)' determined on the basis of a site-specific risk assessment and adopted as site-specific clean-up requirements

Risk-based Criteria. Dissatisfaction with these historical approaches to environmental criteria development[16] is now leading to the establishment of risk-based environmental assessment criteria for soils and groundwaters that combine the simple screening function of numerical criteria within a tiered risk-based rationale. One example of this approach is the American Society for Testing and Materials' (ASTM) guide for 'Risk-Based Corrective Action at Petroleum Release Sites' (RBCA).[17] Here, the first tier of environmental criteria

[16] S. E. Hrudey and S. J. Pollard, *Environ. Rev.*, 1993, **1**, 55.
[17] American Society for Testing and Materials (ASTM), *Emergency Standard Guide for Risk-Based Corrective Action Applied at Petroleum Release Sites*, ASTM, Philadelphia, 1994.

13

(air, soil and groundwater concentrations) for screening petroleum-contaminated sites are a set of risk-based screening levels. Sites with media concentrations in excess of these levels require higher levels of site characterization with site-specific second and third tier criteria being developed to achieve accepted *de minimis* levels of risk.

This general approach has also been adopted by the Dutch Government in the 1994 Soil Protection Act, with the exposure assessment model 'CSOIL' being used to derive new, risk-based intervention values for soil and groundwater at contaminated sites.[18] The UK DoE strategy has also been similar, with the 'CLEA' exposure assessment model being used to derive new, risk-based criteria to replace the ICRCL trigger concentrations. A useful discussion of the UK approach is provided in the Royal Commission on Environmental Pollution (RCEP) 19th Report on the Sustainable Use of Soil.[19]

Notwithstanding these improvements to the regulatory approach, problems remain with the limited ability of measurements of total concentrations to reflect the different behaviours and toxicities of combined or dissociated forms of contaminants, or of isomers of individual contaminants within soil.[16] Likewise, continued use of criteria based on solvent-extractable matter such as toluene-extractable matter, as surrogates for the organic characterization of contaminated soils, ignores the different solubilities of organic compounds across a wide range of solvent polarities. Analytical efforts in such cases need to progress beyond a rudimentary screen of solvent-extractable matter if they are to characterize contaminant-specific risks.[20]

Recent Regulatory Framework

The contaminated land provisions in the UK introduced in the 1995 Environment Act insert a new Part IIA into the Environmental Protection Act 1990 (EPA 1990) consisting of sections 78A to 78YC. The background to the legislation and a summary of the provisions are available.[21] In brief, the Act provides a new regime for the control of threats to health and the environment from land contamination and enacts the policy objectives set out in the UK Government's 1994 paper 'Framework for Contaminated Land' (Department of the Environment, 1994).[8] The key features of the new legislation are:

- a new definition of contaminated land
- endorsement of the 'suitable for use' approach: the new regime deals with unacceptable risks that arise from the current use of the land; change of use issues continue to be addressed by the existing planning process

[18] F. A. Swartjes and R. van den Berg, *Remediation of Contaminated Soil and Groundwater: Proposals for Criteria and Priority Setting*, Royal Institute of Technology Workshop on Contaminated Soils, Department of Chemical Engineering, Stockholm, 1993.

[19] Royal Commission on Environmental Pollution (RCEP), *Sustainable Use of Soil*, 19th Report of the RCEP, HMSO, London, 1996, pp. 130–133.

[20] S. J. T. Pollard, S. L. Kenefick, S. E. Hrudey, B. J. Fuhr, L. R. Holloway and M. Rawluk, in *Analysis of Soil Contaminated with Petroleum Constituents*, ed. T. O'Shay and K. B. Hoddinott, ASTM, Philadelphia, 1994, pp. 39–52.

[21] O. McIntyre, *J. Property Dev.*, 1996, **1**, 5.

- the primary regulatory role rests with the local authorities, reflecting their existing role with respect to statutory nuisance and as planning authorities; the Environment Agencies will act as the enforcing authority for a new designation of 'special' sites
- both enforcing authorities are required to keep a public register of regulatory action
- contaminated land is to be identified on the basis of a risk assessment that uses the source–pathway–receptor approach in the first instance
- where feasible, the 'polluter pays' principle will be applied and persons who caused or knowingly permitted contamination will bear the responsibility for paying for remediation required by a remediation notice; where the polluter cannot be found, responsibility will pass on to the owner/occupier
- the Environment Agencies will provide local authorities with site-specific guidance, regulate special sites, publish periodic reports on the status of contaminated land, sponsor technical research and act as centres of expertise on the subject

The provisions are currently anticipated to come into effect in 1997 and are regarded as addressing much of the uncertainty that previously existed over issues of land contamination. They aim to ensure that actions on contaminated land are directly linked to a technically well-founded assessment of risk. Land will be determined as contaminated only where connectivity between the source, pathway and receptor is established and there is at least one of the following:

 (i) evidence of significant harm, or
 (ii) the significant possibility of significant harm, or
 (iii) the pollution of controlled waters, or
 (iv) the likelihood of pollution of controlled waters

In this manner, the perceived automatic blight associated with the 'potentially contaminative use' approach put forward by Section 143 (EPA 1990) registers is addressed, action is prioritized on unacceptable actual or potential risks to human health and the environment, and sites are improved as and when hazards need to be dealt with.[8]

Application to the Water Environment

The Environment Agency in England and Wales is empowered to take 'anti-pollution works' under section 161 of the Water Resources Act 1991 to deal with the polluting effect of contaminated land on controlled waters and recover the costs from the responsible person. The Scottish Environmental Protection Agency (SEPA) has similar powers under sections 46A–C of the Control of Pollution Act 1974, as subsequently amended.

'Pollution of controlled waters' is defined in the Environment Act 1995 contaminated land provisions as meaning the entry into controlled waters of any poisonous, noxious or polluting matter or any solid waste matter. In their implementation of the new regime, local authorities are to apply a similar approach to that currently used by the Agencies in determining whether the pollution of controlled waters is, or is likely to be, caused. In the latter case, these approaches place a strong emphasis on determining contaminant concentration and availability (combined form, inherent aqueous solubility), mobility (soil adsorptive capacity, porosity and permeability, effective rainfall), the presence of intervening sub-strata, and the presence or absence of existing risk management measures. Determination of these factors relies in part on the results of appropriate leaching tests that establish the propensity of *in situ* contaminants to move through the soil. The importance of these tests is likely to increase substantially under the new contaminated land regime.

Uncertainty Issues

Implementation of the new contaminated land regime in the UK will do much to address the uncertainties associated with the subject in the sense of providing a consolidated piece of legislation and defining responsibilities to deal with the issue. Site-specific uncertainties, however, will always be present, given the complexity of the soil environment. These need to be acknowledged at the outset of an investigation so that effort and resources are focused on issues that improve the overall quality of the information gathered in support of the risk assessment.

Analytical precision and accuracy are issues that require continual scrutiny throughout site assessment and remediation, and many of these issues are now being addressed through various inter-laboratory proficiency schemes. A similar level of attention must be paid to the representativeness of sampling methodologies and sampling techniques. QRA methodologies are constrained by the quality of the baseline toxicological information on which they rely and the relevance of the numerous assumptions they employ in the preparation of risk estimates. Acknowledging such uncertainties is not to negate analytical chemistry or risk science but allows appropriate use of the output from these disciplines for the particular audience, whether a regulator, land owner, financier or member of the public.

4   An Approach to Site Investigation

Site investigation provides the means by which the basic data blocks for risk assessment are assembled. When designing an investigation, it should always be remembered that the outputs must serve several purposes:

- definition of sources of contamination
    location of contaminant
    nature of contaminant
    concentration
    total loading
- identification of pathways
    site topography
    soil/rock permeability

joint/bedding systems
man-made pathways (shafts, pipe backfill, *etc.*)
surface drainage channels
- location of sensitive receptors
depth to groundwater
proximity of surface water

At the same time as determining data for contamination assessment, it may also be appropriate to consider geotechnical issues, such as ground stability, which could affect the nature of remediation measures.

Investigations are most effective when phased. The most common phasing consists of the following steps:

(i) desk study
(ii) main intrusive investigation
(iii) supplementary intrusive investigation

Monitoring of piezometers and standpipes can take place in parallel with intrusive investigations, and the benefit of monitoring over a period of time should not be underestimated. Figure 2 sets out the framework for undertaking investigations of potentially contaminated land.

Desk studies will include historical research of a range of sources to obtain the maximum available information on the site and the previous industrial uses. Sources may include Ordnance Survey maps, previous site owners, regulatory authorities such as the Environment Agency, landfill operators and planning and environmental health authorities. In addition; geological, hydrological and hydrogeological records should be searched.

On completion of the desk study, the main investigations can be designed, either as a single comprehensive survey or as a series of stages targeting specific high-risk areas or problems. The investigation may use a range of techniques, some of which involve non-intrusive activities and others which require physical sampling on site. These techniques are summarized below:

- Non-intrusive
  site walkover
  surface gas emission testing
  geophysical testing
  false colour infrared photography
  thermography
  tracer gas testing
- Intrusive
  boreholes
  trial pits and trenches
  probing techniques
  static cone penetrometer
  window sampling
  gas and water monitoring wells

Figure 2 Investigation process for contaminated land

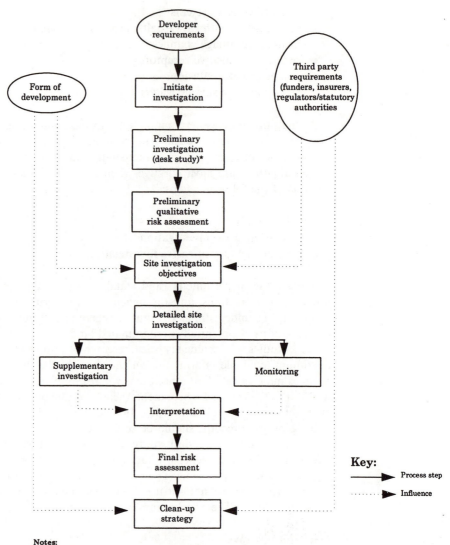

Notes:
* If the desk study shows no evidence of possible contamination, then the process may be terminated. However, investigation of the site for geotechnical purposes should also consider the possibility of contamination, irrespective of the outcome of the desk study.

All opportunities for data collection should be seized upon, and data should be collated and assessed in a systematic manner. For example, when installing monitoring wells, the borehole should be fully logged and described to provide a complete record of the strata. This will aid both decisions about the depth to which screen sections should be installed, and a wider understanding of the site conditions and potential migration pathways.

The extent of investigation of a site is very much a matter for professional judgement based upon an understanding of the history of site usage (derived from the desk study) and the proposed future use. The degree of confidence which can be placed upon the results of an investigation are a function of the following variables:

- the number of soil, water and gas samples taken
- the number of samples tested
- the number of exploratory holes from which the samples are taken
- the spatial layout of exploratory holes
- the extent of any *in situ* testing
- the frequency and duration of subsequent monitoring

Whilst it would be easy to design an investigation with a very large number of sampling and testing locations, this would be a costly and inefficient approach, and investigations should always seek to balance the degree of confidence required against the total cost of the work. Much work is being carried out on sampling strategies[22] and this provides a rational basis for designing investigations.

5   Remediation Strategy

If risk assessment establishes that there is an unacceptable risk of a pollution event or hazard exposure to a sensitive target, then it will be advisable to undertake remedial measures or change operational practices to minimize future risk. In the context of contaminated land, this may mean carrying out additional work during redevelopment to achieve one or more of the following objectives:

- eliminate the hazard by total removal/treatment
- control the hazard by encapsulation or separation
- cut off pathways for contaminant movements by use of barriers or sealing systems
- remove or protect sensitive targets

If redevelopment is not proposed, it may still be necessary to carry out work on land to guard against future liability, and this may be particularly the case for sites which have ongoing active industrial use. Furthermore, if the work is to be undertaken without any prospect of enhanced value, then a further stage to risk assessment is sometimes considered which compares the magnitude and likelihood of financial penalties against the cost of clean-up. This balance is not just a financial one, however, as legal penalties associated with environmental damage in recent legislation also include prison terms and disqualification of directors. Thus any person considering just the financial aspects of damage to the environment should also take legal advice on the additional risks that he or she runs.

When determining the best course of action for remedial/control works, the following factors must be taken into account:

- engineering feasibility
- economics
- health and safety

---

[22] Department of the Environment (DoE), *Sampling Strategies for Contaminated Land*, The Centre for Research into the Built Environment, The Nottingham Trent University, CLR4, DoE, London, 1994.

- environmental impact
- current and future standards/codes of practice
- legislation

Engineering feasibility must be considered for any physical solution which is deemed necessary at a contaminated site. This means that the solution must be practical and buildable, and appropriate to the ground conditions. Schemes which require separation of contaminated soil from clean when the two are visually indistinguishable must be designed with the time delays of chemical testing in mind. Cover systems should combine natural and synthetic layers as appropriate, and the thickness of layers must take into account both their theoretical performance and the practicalities of laying specified materials to a given thickness.

Economics are fundamental to the progression of a scheme, and having established through the risk assessment that there is a requirement for clean-up, then cost must be controlled to ensure that the development scheme can generate sufficient funds to cover the additional reclamation costs. This condition applies for both the public and private sector.

When designing the clean-up strategy, both health and safety and environmental impact issues must be considered such that when a solution is finally chosen, it can be demonstrated to have considered all issues in addition to basic cost. The chosen solution should balance all environmental impacts against cost and final site condition to provide the required reduction in risk, the minimization of environmental impacts and a reasonable cost, thus following the strategy of the Environmental Protection Act of Best Available Technology Not Entailing Excessive Cost (BATNEEC). Furthermore, the Construction Design and Management (CDM) Regulations (1994) require that hazards to the construction workforce are addressed by the Client, the Designer and the Principal Contractor, and this is particularly an issue for contaminated land.

Techniques for dealing with contaminated land in the UK have evolved over the last 30 years from relatively cosmetic solutions of thin soil covers, designed to support minimal vegetation, through to more sophisticated treatment technologies or engineered encapsulation solutions. There are three broad categories of solution which are currently available:

- risk avoidance methods
- engineering methods
- process-based methods

Risk avoidance options include changing land use, site layout and location of services and other infrastructure, thus reducing the risks to particular targets through changing the pathway and isolation of the target. They are not usually feasible where development already exists and are limited to the type of receptor they will protect. Furthermore, if a particular site is designated for, say, housing under a Local Plan, then a change of use may be very difficult to achieve. Assuming that all viable risk avoidance techniques have been adopted, then either engineering or process-based methods or a combination of both must be adopted to deal with the remaining risk.

Engineering methods are based upon conventional civil engineering techniques and are used to remove or isolate the sources of contamination or modify the pathway. They broadly comprise:

- excavation (may be followed by off-site or on-site disposal)
- in-ground barriers such as cut off/vent trenches and slurry walls
- covering and capping systems
- hydraulic measures such as pumped control of contaminated groundwater

Whereas engineering methods do not alter the state of contamination, process-based methods may be defined as treatment to remove, stabilize or destroy contaminants. The unit processes developed for cleaning contaminated soils can be divided into categories, depending on their general operating principles. The categorization is derived from work into waste treatment processes and is as follows:

- biological
- chemical
- physical
- solidification
- thermal

When considering a remediation strategy for a site, six factors have been highlighted as requiring consideration. Of these six, economics is generally the deciding factor when other issues have been taken into account and perhaps two or three options are left. One option which entails twice the expenditure of another is unlikely to be taken up unless the benefits of the more expensive option can be seen to be worth the increased cost. For example, capping contaminated soil with engineered layers of clean material may be feasible, have minimal impact on the environment, accord with Codes of Practice and legislation and not create any health and safety problems. However, the contamination will remain on site, and some parties such as banks or funding institutions may view this as unacceptable and require total elimination of all contaminants from site, either by off-site landfill or treatment. A much higher cost may attach to this second option which reflects this 'whiter than white' approach, but this higher cost may tip the scales in terms of the economics of the overall development.

Whatever solution is chosen, it is absolutely essential to ensure a high standard of implementation; this may be done *via* both the normal controls of a construction contract and the quality assurance procedures which most companies now embrace British and European Quality Standards. The documentation of the works is important to demonstrate to future site owners and tenants or those who fund purchase of the land that the remediation has been successful in meeting targets and objectives set. Verification of the remediation *via* a systematic sampling and testing programme is often the best way of demonstrating the achievement of the set objectives.

6 Use of Consultants

The role and performance of consultants is increasingly important in delivering technically sound decisions in contaminated land management. The development of an optimum relationship between client and consultant has been hampered in the past by variable quality in the market place. However, there is strong evidence that access to appropriate consultancy advice is now a critical factor in business becoming comfortable that the challenges of contaminated sites have technical and affordable solutions.

At present, no professional institution can properly lay claim to contaminated land as its special sphere of interest. Good consultancy in contaminated land requires inputs from a number of separate disciplines. It is unlikely that a single individual consultant would ever have all the detailed professional knowledge, expertise and technical skills to be able to deal with the more complex situations that can arise regarding contaminated sites. However, a good appreciation of the concerns of other specialists/professionals will usually be required, and is of increasing importance as consultants assume greater responsibility for project specification and management.

Historically, consultants dealing with contaminated land have come from a range of disciplines and backgrounds. This is likely to remain the case, and should be welcomed because of the multidisciplinary and complex nature of the problems that may be encountered. A number of organizations have been formed to assist in communication of cross-discipline issues. Examples include the Environmental Consultancy Group (ECG) of the Environmental Industries Commission (EIC), formerly the Association of Environmental Consultancies (AEC), the Association of Geotechnical and Geoenvironmental Specialists (AGS), and the Forum on Contamination in Land (FOCIL) (Table 6). These groups have a key role in telling clients who wish to use consultants what is on offer and how to engage specialist advice in a cost-effective and beneficial way.

An example of the information now being produced to explain and define the role of consultancy in contaminated land is the forthcoming DoE report on Quality Assurance in Environmental Consultancy.[23] This addresses the responsibilities and expectations of a competent consultant practising in the field of contaminated land. In addition, many organizations with practical or legal responsibilities for contaminated land have also been developing guidance to improve the realization of expectations, including the usual circumstances when consultancy support is involved. Examples include the Institute of Environmental Health Officers,[24] the National Federation of Housing Associations,[25] Scottish Enterprise,[26,27] the Welsh Development

---

[23] *Quality Assurance in Environmental Consultancy: A Quality Approach for Contaminated Land Consultancy*, report by Association of Environmental Consultancies in Association with the Laboratory of the Government Chemist, DoE, London, in preparation.

[24] Institution of Environmental Health Officers (IEHO), *Contaminated Land: Development of Contaminated Land—Professional Guidance*, IEHO, London, 1989.

[25] National Federation of Housing Associations (NFHA), *Contaminated Land: Issues for Housing Associations*, NFHA, London, 1995.

[26] Scottish Enterprise (SE), *How to Investigate Contaminated Land: Requirements for Contaminated Land Site Investigations*, SE, Glasgow, 1994.

Table 6 Contact points for advice in consultancy capabilities in contaminated land	**The Association of Geotechnical and Geoenvironmental Specialists** *39 Upper Elmers End Road* *Beckenham* *Kent* *BR3 3QY*	**The Environmental Industries Commission** *6 Donaldson Road* *London* *NW6 6NB*
	Environmental Contacts: A Guide for Business *DTI Environmental Publications* *ADMAIL 528* *London* *SW1W 8YT*	**Environment Industry Yearbook** *Environment Press Ltd* *26 Brock Street* *Bath* *BA1 2LN*
	Environment Business *Information for Industry Ltd* *521 Old York Road* *London* *SW18 1TG*	**The Forum on Contamination in Land (FOCIL)** *c/o Royal Institute of Chartered Surveyors* *12 Great George Street* *Parliament Square* *London* *SW1P 3AD*

Agency[28] and the Construction Industry Research and Information Association.[29]

It can be concluded that the sophistication of and understanding between client and consultant has changed and improved rapidly, in response to the technical and legal development of the subject area. Very few organizations which own, develop or administer contaminated land have the expertise to address all foreseeable issues adequately. Indeed, efficient contaminated land work must rely on such a breadth of professional and technical experience it is rarely cost effective for organizations to be self-sufficient. Those with the greatest track record in dealing with historical legacies of contamination, for example in the chemicals, oil and gas, waste and metal industries, are the largest users of specialist consultancy support.

The essence of effective consulting support is quality assured and clearly presented reports and documentation. Even where specialist investigation or remedial techniques are being implemented, the value of the site work is dependent on effective reporting of findings and actions carried out. Many third parties, whether they be regulator, developer, purchaser, auditor or the local public, will require to understand what has been done to manage a contaminated

[27] Scottish Enterprise (SE), *How to Approach Contaminated Land: A Framework for the Assessment of Contaminated Land and Selection of Remedial Options*, SE, Glasgow, 1994.

[28] Welsh Development Agency (WDA), *Manual on Remediation of Contaminated Land*, WDA, Cardiff, 1993.

[29] Construction Industry Research and Information Association (CIRIA), *Remedial Treatment for Contaminated Land*, SP101–SP111, CIRIA, London, 1995, vols. I–XI.

Figure 3 Consultant reports for site-based projects[23]

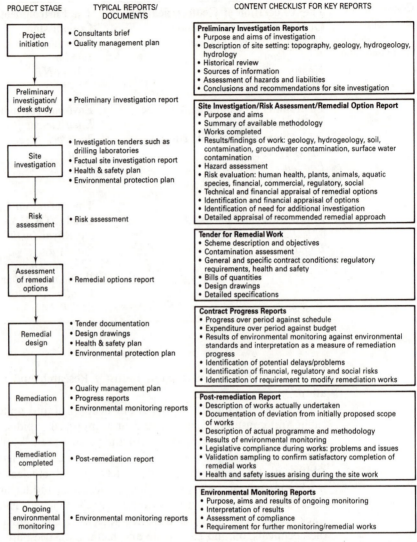

PROJECT STAGE	TYPICAL REPORTS/ DOCUMENTS	CONTENT CHECKLIST FOR KEY REPORTS
Project initiation	• Consultants brief • Quality management plan	**Preliminary Investigation Reports** • Purpose and aims of investigation • Description of site setting: topography, geology, hydrogeology, hydrology • Historical review • Sources of information • Assessment of hazards and liabilities • Conclusions and recommendations for site investigation
Preliminary investigation/ desk study	• Preliminary investigation report	**Site Investigation/Risk Assessment/Remedial Option Report** • Purpose and aims • Summary of available methodology • Works completed • Results/findings of work: geology, hydrogeology, soil, contamination, groundwater contamination, surface water contamination • Hazard assessment • Risk evaluation: human health, plants, animals, aquatic species, financial, commercial, regulatory, social • Technical and financial appraisal of remedial options • Identification and financial appraisal of options • Identification of need for additional investigation • Detailed appraisal of recommended remedial approach
Site investigation	• Investigation tenders such as drilling laboratories • Factual site investigation report • Health & safety plan • Environmental protection plan	
Risk assessment	• Risk assessment	
Assessment of remedial options	• Remedial options report	**Tender for Remedial Work** • Scheme description and objectives • Contamination assessment • General and specific contract conditions: regulatory requirements, health and safety • Bills of quantities • Design drawings • Detailed specifications
Remedial design	• Tender documentation • Design drawings • Health & safety plan • Environmental protection plan	**Contract Progress Reports** • Progress over period against schedule • Expenditure over period against budget • Results of environmental monitoring against environmental standards and interpretation as a measure of remediation progress • Identification of potential delays/problems • Identification of financial, regulatory and social risks • Identification of requirement to modify remediation works
Remediation	• Quality management plan • Progress reports • Environmental monitoring reports	**Post-remediation Report** • Description of works actually undertaken • Documentation of deviation from initially proposed scope of works • Description of actual programme and methodology • Results of environmental monitoring • Legislative compliance during works: problems and issues • Validation sampling to confirm satisfactory completion of remedial works • Health and safety issues arising during the site work
Remediation completed	• Post-remediation report	
Ongoing environmental monitoring	• Environmental monitoring reports	**Environmental Monitoring Reports** • Purpose, aims and results of ongoing monitoring • Interpretation of results • Assessment of compliance • Requirement for further monitoring/remedial works

site. Usually, these requirements are met by appropriate reports. Confidence in advice and action is dependent on good communication. Figure 3 lists the reports which are characteristic of site-based projects. The following articles aim to provide more detail on the processes and techniques which are applied to deliver such site-based projects from initial data gathering through risk assessment to successful implementation of all necessary remediation.

UK Government Policy and Controls

MARY R. HARRIS AND JUDITH DENNER

1 Introduction

To the casual observer, it must seem as though 'contaminated land' did not exist in the UK until 1991 when the Government first issued proposals for the registration of land which may be contaminated. For many organizations, the debate which followed publication of the Government's proposals shifted contaminated land from a position of relative obscurity to a conspicuous place at the top of the environmental agenda. In fact, the UK was one of the first advanced industrial nations to realize that the legacy of industrial development had implications beyond those of simple physical dereliction; government guidance on the identification and use of contaminated land had been in force for some time when the registration proposals were issued, and a substantial body of practical experience in redeveloping contaminated land already existed.

The Government's registration proposals were a direct response to recommendations made by a Select Committee Inquiry into contaminated land in 1990.[1] This has since proved to be an important marker for change in the pace of UK policy development. However, the proposals also coincided with increased international scrutiny of the much wider issues of environmental damage, how and to what extent it should be rectified, and who should pay. The Government's registration proposals met with strong opposition because land owners feared that if the proposals were implemented they would cause blight. In response, in March 1993, the Government withdrew its proposals and set up a wide ranging policy review on controlling contaminated land and meeting the cost of remedying damage to the environment. The review included public consultation on a paper entitled *Paying for Our Past* and concluded in November 1994 with the publication of the policy statement *Framework for Contaminated Land*. This looked forward to the Government introducing a modern contaminated land power which was subsequently put on the statute book in the form of the Environment Act 1995. In parallel with these changes in policy and legislation, there has been increased investment in technical research and development, and greater international exchange on both the policy and technical fronts.

There is no doubt that contaminated land is an important issue. Evidence of a

---

[1] House of Commons Select Committee on the Environment, *First Report on Contaminated Land*, HMSO, London, 1990, vols. 1–3.

substantial, but often environmentally damaging, industrial past is obvious to all those involved in the assessment of land in many areas of the UK. It is also a complex issue, with some technical uncertainties still to be resolved, and it can be expensive to treat. Contaminated land is not an insurmountable problem—there are well established ways in which uncertainties can be managed, complexities reduced and costs controlled—but it does require sometimes difficult choices to be made. At the highest level, society chooses what priority to give to environmental protection in general, and soil and water quality in particular. Society also decides the balance to be struck between establishing and enforcing high standards of environmental protection, and maintaining sufficient wealth creating capacity to invest in environmental improvement. Even at the lowest level of an individual site, important choices have to be made: between the cost of collecting information about a site and the benefits which such information may bring in terms of increased technical confidence.

In practice, the key advances made by the UK over recent years have been to create better conditions for well informed decision-making, rather than any striking breakthrough in science and technology. At the very least, the last five years have seen wider participation in the policy debate. Significant improvements in the scope, content and presentation of UK technical guidance should ensure a more consistent and reliable technical output, and hence better informed financial and legal decision-making. The UK's participation in international initiatives should encourage an early consensus on the best way of managing contaminated land.

To a large extent, agreement on the most effective technical approach is already in place. The application of conventional risk assessment and risk management principles already provides a structure around which different stakeholders can develop their own particular strategies for managing contaminated land. There is evidence that industry, landowners, developers, regulators, financial and insurance organizations, and the professional institutions, are taking steps along these lines. The next five years are likely to see further development and consolidation in this area. Meanwhile, new monitoring and reporting duties on local authorities and the Environment Agencies in England and Wales, and in Scotland, should generate the first national statistics on the incidence and treatment of land which is presenting, or is likely to present, unacceptable risks to health or the environment.

2 Development of UK Policy

Key Developments

Key developments in UK contaminated land policy and legislation over the last 20 years are shown in Table 1.

First Recognition of the Problem and Establishment of the ICRCL. In the mid-1970s, several local authorities reported that they were experiencing problems in redeveloping industrial land which had become contaminated as a result of previous uses.[2] In an effort to provide a co-ordinated response to these

---

[2] M. R. Harris, in *Reclaiming Contaminated Land*, ed. T. C. Cairney, Blackie, Glasgow, 1987.

Mid 1970s	First recognition of the problem of redeveloping land previously used for industrial and related purposes
Early 1980s	Establishment of an interdepartmental advisory body (ICRCL)
Mid to late 1980s	Development of technical guidance on contaminated land
1990	House of Commons Select Committee on the Environment Inquiry on contaminated land
1990	Environmental Protection Act 1990
1991–1992	Consultation on proposals for the registration of potentially contaminated land based on past and current land use
1993	Initiation of Government review of contaminated land and liability
1993	Publication of the European Commission Green Paper on Environmental Liability
March 1994	Publication of consultation paper, *Paying for Our Past*, as part of the review
November 1994	Conclusion of Government review and publication of UK policy paper, *Framework for Contaminated Land*
1995	Environment Act 1995 (Part IIA of EPA 1990)
1996	Draft statutory guidance

difficulties, and a centralized source of information and guidance, the Government established the Interdepartmental Committee for the Redevelopment of Contaminated Land (ICRCL). Several Government departments were represented on ICRCL, including the Department of the Environment, Department of Health, Welsh Office, Scottish Office, Ministry of Agriculture, Fisheries and Food, and the Health and Safety Executive. The early objectives of the Committee were to increase awareness of the potential problems associated with the redevelopment of contaminated land, produce technical guidance and promote research.[3] An important output of the Committee was a series of Guidance Notes covering different types of industrial land, and on the assessment of contaminated land intended for development.[4]

Several other initiatives were taken in the mid to late 1980s to ensure that contaminated land was identified well in advance of any action to redevelop it for another use, and that appropriate measures were in place to ensure safe redevelopment. Thus in 1985 the Building Regulations were modified to ensure the potential for hazardous substances in the ground was taken into account in building projects,[5] and in 1987, Government guidance to local planning

---

[3] ICRCL, *Progress Report of the Interdepartmental Committee on the Redevelopment of Contaminated Land*, ICRCL Paper 19/79, DoE, CDEP, London, 1979.

[4] ICRCL, *Guidance on the Assessment and Redevelopment of Contaminated Land*, ICRCL Guidance Note 59/83, 2nd edn., DoE, London, 1987.

[5] Department of the Environment & Welsh Office, *Site Preparation and Resistance to Moisture*, Approved Document C, Building Regulations 1985, HMSO, London, 1985 [note this guidance has been superseded, see Ref. 34].

authorities advised that the presence of, or potential for, contamination was a 'material consideration' for planning purposes.[6] In 1988, the British Standards Institution published a Draft for Development Code of Practice on the identification and investigation of potentially contaminated land.[7]

House of Commons Select Committee Inquiry. In 1990 the House of Commons Select Committee on the Environment held an inquiry on contaminated land following their inquiry on hazardous waste which had first identified contaminated land as an issue. The 1990 inquiry report, based on both oral and written submissions, was heavily critical of certain aspects of UK Government policy in this area. The main criticisms were:

● The narrow working definition of contaminated land which referred only to that land which is contaminated and 'potentially available for development', thus apparently excluding other potential categories of contaminated land, including land already in use and land affected by the migration of contaminants.
● The lack of reliable information on the scale, nature and distribution of contaminated land in the UK.
● The failure to encourage active consideration of the wider environmental protection and pollution control aspects of land contamination because of the prominence given to redevelopment, although the Committee broadly endorsed the UK approach of taking the proposed end-use into account when redeveloping contaminated land.
● Limitations in the available technical guidance, particularly in relation to the water environment and the range of contaminants covered.
● The failure to encourage the use of a broad range of remedial techniques, other than conventional containment and off-site disposal, through research and development, improved technical guidance and better-targeted grant assistance.

The Committee produced 29 separate recommendations for consideration by the Government. Some called for the introduction of specific legislation, others for reorganization of Departmental responsibilities for policy development and the provision of guidance. Many highlighted the general need for a more comprehensive policy approach.

The Government published its response to the Committee's report in July 1990.[8] It did not agree with all the conclusions drawn by the Committee, nor did it accept that action was required in response to all of the Committee's

---

[6] Department of the Environment & Welsh Office, *Development of Contaminated Land*, Circulars 21/87 and 22/87, DoE, London, and Cardiff, 1987 [note this guidance is now superseded by Planning Policy Guidance Note No. 23, HMSO, London, 1994].

[7] British Standards Institution, *Draft for Development Code of Practice on the Identification of Potentially Contaminated Land and its Investigation*, DD175, BSI, London, 1988.

[8] Department of the Environment, *Contaminated Land: The Government's Response to the First Report from the House of Commons Select Committee on the Environment*, HMSO, London, 1990.

recommendations. The Government did concede, however, that more could be done to improve information on the nature and scale of contaminated land, on assessment and in the area of research and development.

Introduction of the EPA 1990.   In 1990 the Environmental Protection Act (EPA) reached the statute book. The legislation was widely regarded as one of the most comprehensive pieces of environmental protection legislation ever to have been introduced in the UK. Amongst other things, the EPA introduced a new regime for the regulation of industrial facilities under Integrated Pollution Control arrangements, as well as a completely new set of controls over waste management activities. The legislation also contained two sections (s143 and s61) with implications for the management of contaminated land.

Section 143 of the EPA 1990.   One of the most important statements made by the Select Committee in 1990 was that it was not convinced that the Department of the Environment's estimates of the scale of contaminated land 'provide a sound factual basis for policy-making in this area'. The Committee recommended that the Government bring forward legislation 'to lay on local authorities a duty to seek out and compile registers of contaminated land'.

The Committee recommended that registers were drawn up on the basis of the actual presence of contamination in land, but the Government accepted the view of an internal working party that it would be simpler if they were to be based on past and present 'contaminative uses' of land. Section 143 of the EPA imposed a duty on local authorities to compile the relevant information. Public consultation on the proposals took place during 1991 and 1992.

The original proposals provided for the registration of sites previously or currently used for a wide range of different contaminative uses, defined as any use of land which may cause it to be contaminated with noxious substances. At the time consultation documents were issued (May 1991 in England and Wales, and August 1991 in Scotland), regulations implementing the registration process were expected to take effect on 1 April 1992. Public access to the registers was planned for one year later, *i.e.* by April 1993.

In March 1992, Ministers announced that there would be a delay in the implementation of Section 143 due to the number of comments received *via* the public consultation process and the need for further consideration of the proposals. It soon became clear that while there was positive support, serious concerns about the proposed registers were being expressed by property owners and the funding institutions in particular, on the grounds of blight and negative effect on property values.

In July 1992 the Department of the Environment issued draft regulations for further comment.[9] These substantially reduced the range of contaminative uses to be brought within the scope of the registration process (at least initially) and proposed a two-tiered approach to registration, depending on whether site investigation or other works had been carried out on the relevant land.

[9] Department of the Environment, *Draft Environmental Protection Act* 1990 (*Section* 143 *Registers*) *Regulations*, Consultation paper, DoE, London, 1992.

Despite these revisions, and in the face of continuing opposition, on 24 March 1993 the Government withdrew its proposals for establishing registers of contaminative uses of land and announced a wide-ranging review of problems in this area.[10]

EC Green Paper on Environmental Damage.   On 17 March 1993 the European Commission issued its Green Paper on repairing damage to the environment.[11] The paper was a general discussion of the usefulness of a civil liability regime for rectifying environmental damage. Although the paper addressed environmental damage in its widest sense, including damage to the unowned environment, much of its content was particularly relevant to the issue of contaminated land.

The paper distinguished between criminal and civil liability regimes, and between fault-based and strict civil liability systems. It also covered:

- the definition of environmental damage
- limits on liability
- defences (including whether those who operate in accordance with Government authorizations or permits could still be held liable if damage to the environment subsequently occurred)
- the identity of those with the necessary standing to initiate legal action
- the role of insurance and joint compensation systems in the event that the responsible party could not be found or was insolvent
- whether any system should be retrospective in application

The Government responded to the issues raised by the EC Green Paper in October 1993, making it clear that it did not share the Commission's view that there was a need to introduce a European-wide system of civil liability for environmental damage.[12] Although definitive conclusions were not drawn on all the issues raised by the Green Paper, the response gave some early indications of the Government's thinking on the more specific issue of the liability which may be associated with contaminated land.

'Paying for Our Past'.   In March 1994 the Government issued an interim consultation paper, *Paying for Our Past*, setting out the preliminary conclusions of the Government review.[13] This paper:

- set out the existing policy and legal framework for contaminated land, and associated potential liabilities

---

[10] Department of the Environment, *Michael Howard Announces Review of Land Pollution Responsibilities*, DoE Press Release No. 209, DoE, London, 24 March 1993.

[11] European Commission, *Green Paper on Remedying Environmental Damage*, Com (93)47, EC, Brussels, May 1993.

[12] Department of the Environment, *Government Response to the EC Green Paper*, DoE, London, October 1993.

[13] Department of the Environment & Welsh Office, *Paying for Our Past*, Consultation Paper, DoE, London, and WO, Cardiff, March 1994.

- discussed the difficulty of developing systems for allocating responsibility for remedying environmental damage that strike the right balance between the interests of site owners, polluters, the public sector and the community at large
- set out the key issues for resolution, and preliminary conclusions on the need for changes or clarification
- invited responses on the issues raised

A separate but parallel consultation exercise was carried out in Scotland, where a Scottish consultation paper, *Contaminated Land: Clean-up and Control*, was issued.[14]

After analysing the results of the consultation, the Government published *Framework for Contaminated Land* in November 1994.[15] This marked the conclusion of the Government's review and paved the way for the contaminated land provisions subsequently incorporated into the Environment Act 1995.

Framework for Contaminated Land

Framework for Contaminated Land reflects many of the views and arguments set out in earlier consultation papers, particularly on the balance the Government is seeking to achieve between maintaining and improving environmental quality and ensuring that businesses and others are able to cope with environmental costs.

The document emphasizes the Government's commitment to two fundamental environmental principles: 'sustainable development' and the 'polluter pays'. It also distinguishes between future pollution and contamination which has arisen from historical activities and events. The document sets out the Government's view that the current legal framework is sufficient both to deter behaviour likely to lead to environmental pollution and to provide remedies in the event that legal requirements are breached. However, it also signals the Government's acceptance that there is a need to deal specifically with historical contamination.

The paper also reaffirms the UK's long-standing policy objective of a 'suitable for use' approach to the control and treatment of existing contamination, arguing that this supports the principle of sustainable development by reducing damage from past activities, allowing land to be kept in, or returned to, beneficial use, and reducing development pressures on greenfield sites.

The importance of development and redevelopment, and the role that both public and private sector organizations can play in tackling historical contamination through the redevelopment process, is also highlighted. The document argues that technical confidence and certainty about the regulatory obligations involved in managing contaminated land are crucial factors in the ability of the market to take voluntary action. It also anticipates the important role likely to be played by the Environment Agency, both as a source of technical guidance and as a regulatory body in its own right.

Risk assessment and risk management concepts are introduced in the Government's statement of its 'suitable for use' approach that, for both development sites within the planning control system and in relation to the

[14] Scottish Office, *Contaminated Land: Clean-up and Control*, SO, Edinburgh, March 1994.
[15] Department of the Environment, *Framework for Contaminated Land*, DoE, London, November 1994.

	Roles and responsibilities	Several identified, including: ● Commercial organizations through the normal development cycle ● Public bodies (*e.g.* English Partnerships, Welsh Development Agency and the Development Corporations), again through development and particularly where contaminated land is contributing towards dereliction ● Regulatory authorities through the enforcement of public health and environmental protection legislation, and the land use planning and development control system
	Powers to take action and a new definition of 'contaminated land'	To reflect existing powers of local authorities (contained in the statutory nuisance provisions of Part III of the EPA 1990) to provide a modern, specific contaminated land power to identify and take action on contaminated land
	Specific arrangements for closed landfill sites	For monitoring closed landfill sites which come within the definition of contaminated land (replacing the unimplemented provisions of section 61 of the EPA 1990)
	Liabilities of polluters and owners/occupiers	Primary responsibility to be carried by polluters in line with the polluter pays principle (PPP) although owners and occupiers may have responsibilities where the original polluter cannot be found or where liability has been transferred through a land/property transaction. Owners or occupiers not to be responsible for the cost of restoring off-site damage which they had not caused or knowingly permitted
	Liabilities of lenders	Lenders not to be entirely relieved of the risks which may be associated with their activities, particularly where they have caused or contributed to damage, but liabilities to be in proportion to responsibilities
	Cost recovery, orphan liabilities and position of homeowners & SMEs	● Powers for enforcing authorities to take action in default and, in some circumstances, to secure their uncovered costs by way of charges on the land ● Recognition that there may be so-called 'orphanage liabilities' where enforcing authorities are unable to recover costs. The Government's objective is to keep such cases to a minimum although not at the expense of pursuing a 'deep pocket' approach where liability is allocated regardless of responsibility ● Regulators to take hardship into account when deciding whether to seek recovery of all, or part, of their costs of taking remedial action; this provision is expected to be of most benefit to individual homeowners and small to medium-sized enterprises (SMEs)

Table 2 Continued

	• Position of individual homeowners and SMEs is also expected to be improved by the increased availability of and access to relevant information held by local authorities and the Agency, and the requirement on enforcement bodies to take hardship into account. *Caveat emptor* to remain in place
Research	The Environment Agency to assume existing DoE responsibilities for research. Particular priority to be given to publishing guideline concentrations for priority contaminants
Burden of proof and cost–benefit analysis	Burden of proof that a person had caused or knowingly permitted the contamination and its effects to rest with the enforcing authority which will also be required to demonstrate that the required action is reasonable given expected costs and environmental benefits
Compilation of information and public access	• The Agency to have an overview role and to be under a duty, from time to time, to take a national overview of, and report on, contaminated land based on information supplied by local authorities • Information on contaminated land held by local authorities and the Agency to be subject to requirements under the Environmental Information Regulations 1992 and those of the Citizen's Charter
s143 of EPA 1990	Section 143 of the EPA 1990 (registers of land that may be contaminated) to be repealed

enforcement activities of regulatory bodies, remedial action should only be applied where:

- contamination poses unacceptable actual or potential risks to health and the environment; and
- there are appropriate and cost-effective means available to do so, taking into account the actual or intended use of the site

Within this overall policy approach, the Government's further policy objectives are stated as:

- to improve sites as and when hazards need to be dealt with, the private sector decides to develop land, or public bodies prepare land to promote development
- to encourage an efficient market in land which may have been contaminated
- to encourage the redevelopment of such land
- to remove unnecessary financial and regulatory burdens

The document goes on to explore in detail the roles and responsibilities of the various parties who may be involved in managing contaminated land, the principles to be considered when allocating liability for damage caused by contaminated land, and the wider functions of both local authorities and the Environment Agency in monitoring and reporting both the incidence of contaminated land and progress in dealing with it. Some of the more important features of the document are set out in Table 2.

Framework for Contaminated Land essentially set the scene for the new contaminated land provisions contained in the Environment Act 1995. However, this legislation is not the only statute relevant to contaminated land. The Act is intended to satisfy an important policy objective, *i.e.* to provide a means by which action can be taken to deal with land which:

- poses unacceptable actual or potential risks to health or the environment such that the land cannot be considered fit for its current use or environmental setting; and which
- cannot be addressed by other means.

The provisions of the Act will not apply to land where contamination is being dealt with through the planning and development control process, or to contamination at operational premises or facilities (such as waste disposal sites) which has resulted from breaches of existing environmental protection legislation, such as the EPA 1990 and Water Resources Act 1991. These items of legislation have an important preventative function, *i.e.* to ensure that regulated facilities or land are handled in such a way that control *via* the Environment Act 1995 is avoided in future.

Government Research and Development Programme

Department of the Environment spending on the contaminated land research programme amounts to about £1.3 million per annum. Key research priorities are listed in Table 3. The output has been published by the Department of the Environment in the form of Contaminated Land Research Reports[16-21] and in a series of Industry Profiles.[22]

The Department of the Environment is also preparing procedural guidance to improve the quality assurance of technical work. One research project is intended

[16] Department of the Environment, Contaminated Land Research Report 1, *A Framework for Assessing the Impact of Contaminated Land on Groundwater and Surface Water*, DoE, London, 1994, vols. 1 and 2.

[17] Department of the Environment, Contaminated Land Research Report 2, *Guidance on Preliminary Site Inspection of Contaminated Land*, DoE, London, 1994, vols. 1 and 2.

[18] Department of the Environment, Contaminated Land Research Report 3, *Documentary Research on Industrial Sites*, DoE, London, 1994.

[19] Department of the Environment, Contaminated Land Research Report 4, *Sampling Strategies for Contaminated Land*, DoE, London, 1994.

[20] Department of the Environment, Contaminated Land Research Report 5, *Information Systems for Land Contamination*, DoE, London, 1994.

[21] Department of the Environment, Contaminated Land Research Report 6, *Prioritisation and Categorisation Procedure for Sites which may be Contaminated*, DoE, London, 1995.

[22] Department of the Environment, Industry Profiles (various titles), Building Research Establishment, Garston, various dates.

to provide relatively simple, clear and concise model procedures on the identification, assessment, treatment and monitoring of contaminated land which will:

- set out a recommended UK approach to the whole process of managing contaminated land (and hence encourage consistency and transparency in technical practice)
- explicitly describe the objectives, scope, steps to be followed and expected outputs of particular phases or tasks (and hence reduce the scope for uncertainty in the technical approach actually adopted)
- provide a reference point for the more detailed guidance already available to technical users (and hence ensure that good technical practice is consistently applied)
- be accessible to a wide range of users, including those who may not have the relevant technical expertise but who are nevertheless obliged to rely on and interpret the technical output of others

The output of the research will include a Framework of Model Procedures, together with individual procedures on Risk Assessment, Selection and Evaluation of Remedial Options and the Documentation of Remedial Action.[23]

The Construction Industry Research and Information Association (CIRIA) has carried out joint collaborative research projects with Government and industry funding and has published a series of technical guidance documents on contaminated land.[24-32]

Other Government-funded research includes work carried out by the Science Research Councils, such as that under the Waste and Pollution Management Programme, and the LINK programme set up by the Department of Trade and Industry to encourage collaborative research by industry and the universities. The UK Government also plays an active role in research on contaminated land at an international level, as explained in more detail in Section 4.

In 1996, the bulk of the Department of the Environment's contaminated land

[23] S. M. Herbert, M. R. Harris and J. Denner, in *Proceedings of Contaminated Land '95*, ed. W. J. van den Brink, R. Bosman and F. Arendt, Kluwer, Dordrecht, 1995, pp. 701–709.

[24] M. R. Harris, S. M. Herbert and M. A. Smith, *Remedial Treatment for Contaminated Land*, vols. 1–12, Special Publications 101–112, CIRIA, London, various dates.

[25] J. E. Steeds, E. Shepherd and D. L. Barry, *A Guide to Safe Working on Contaminated Sites*, Report 132, CIRIA, London, 1995.

[26] Building Research Establishment (BRE), *Methane and Associated Hazards to Construction: A Bibliography*, Special Publication 79, CIRIA, London, 1992.

[27] P. J. Hooker and M. P. Bannon, *Methane: its Occurrence and Hazards in Construction*, Report 130, CIRIA & British Geological Survey, London, 1993.

[28] D. Crowhurst and S. J. Manchester, *The Measurement of Methane and Other Gases from the Ground*, Report 131, CIRIA, Fire Research Station, BRE, London, 1993.

[29] S. P. Rowan and J. G. Raybould, *Methane Investigation Strategies*, Report 150, CIRIA, London, 1996.

[30] C. R. Harries, J. M. McEntee and P. J. Witherington, *Interpreting Measurements of Gas in the Ground*, Report 151, CIRIA, London, 1996.

[31] G. B. Card, *Protecting Development from Methane*, Report 149, CIRIA, London, 1996.

[32] N. J. O'Riordan and C. J. Milloy, *Risk Assessment for Methane and Other Gases from the Ground*, Report 152, CIRIA, London, 1996.

Table 3 Current DoE research priorities on contaminated land

Information requirements, including the information needed to measure the effectiveness of policy implementation
Risk assessment, including development of guideline values for contaminants affecting human health, collation of toxicological data, prediction of contaminant behaviour, and ecological risk assessment
Analytical techniques, including effectiveness of monitoring technology
Remedial treatment, including established and innovative engineering-based and process-based technologies and techniques
Quality assurance, including a review of quality assurance issues in the assessment of contaminated land
Review of research on contaminated land in the UK

research programme was transferred to the Environment Agency where its main themes are expected to continue, to support the objectives of the Agency in this area.

3 The Legal Framework

A wide variety of different legislation is potentially relevant to the management of contaminated land in the UK, parts of which have been in force for some time. The relevant controls may be exercised through statutory legislation or through the civil law. Although compliance with the statutory legislation will often be the major consideration, in some circumstances civil provisions may be important.

The following sections provide a summary account of the main statutory provisions currently applying in England and Wales. More detailed information on statutory requirements, on parallel legislation in Scotland and Northern Ireland and on relevant civil laws is available elsewhere.[33]

General Controls

Land Use Planning and Development Control. The land use planning and development control legislation may be relevant in a number of important respects, including:

- policy development on the use of land at national, regional and local levels
- the prior approval of 'development', including that where actual or potential contamination is a 'material consideration'
- identifying and minimizing the potential environmental impacts which may be associated with certain types of development.

Potentially relevant provisions are summarized in Table 4.

The Building Act 1984 and Building Regulations 1991 may also be relevant where it is intended to construct buildings on land containing hazardous materials. The Building Regulations require measures to be taken to ensure the reasonable standards of health and safety, welfare and convenience of persons

[33] M. R. Harris, S. M. Herbert and M. A. Smith, *Remedial Treatment for Contaminated Land*, vol. 12, Policy and Legislation, Special Publication 112, CIRIA, London, in press.

Table 4 Main relevant provisions of the land use and development control system	Development of land use policies	Town & Country Planning Act 1990; Planning & Compensation Act 1991	*Via* Statutory Development Plans and in accordance with national planning policy guidance, including that relating to the return of derelict and contaminated land to beneficial use, and in relation to contamination as a 'material consideration'
	Definition of development	TCPA (1990)	To include operations carried out in connection with the land or use of the land including, where appropriate, large-scale site investigation works, remedial works, redevelopment schemes
	Development control	TCPA (1990); Planning & Compensation Act 1991	For example, through conditional planning permissions, planning agreements or obligations
	Enforcement of planning controls	TCPA (1990); Planning & Compensation Act 1991	For example, through Enforcement Notices, Stop Notices, Breach of Planning Conditions Notices, Injunctive Relief
	Consideration of potential environmental impacts	Town & Country Planning (Assessment of Environmental Effects) Regulations 1988	Mandatory in relation to certain forms of development and discretional in certain other cases

using or in the proximity of buildings or otherwise connected with them. The Building Regulations generally apply to new construction but they may be relevant to the extension or modification of existing buildings which may not be addressed by the planning system. Building works are required to comply with Approved Document C which specifically addresses hazardous substances, such as radon and methane, which may be present in, on or under the ground.[34] Approved Document C also prohibits the use of materials in the construction of buildings which may in themselves present a hazard.

Occupational Health and Safety. The management of contaminated land may expose individuals to physical, chemical and biological hazards. Relevant activities include the clearance of redundant buildings and plant, site investigation and remediation works, and the handling and transport of hazardous materials. All such work activities come under the jurisdiction of the Health and Safety at

[34] Department of the Environment & Welsh Office, *Site Preparation and Resistance to Moisture*, Approved Document C, Building Regulations 1991, HMSO, London, 1992; see also Building Research Establishment, *Construction of New Buildings on Gas Contaminated Ground*, BRE, 1991.

Work, *etc.*, Act 1974 and its associated regulations. The Health and Safety at Work, *etc.*, Act places a general duty on employers to protect the health and safety of their employees and other persons (including members of the general public) who may be at risk of harm as a result of hazards arising at a place of work.

A number of different regulations may be relevant in any particular case. Some apply to the management of health and safety matters, others to physical hazards and some to specific hazardous substances. Some of the main potentially relevant regulations are summarized in Table 5.

Public Health Protection. Until the Environment Act 1995 comes into force, the potential public health implications of contaminated land in a non-occupational context is covered under 'nuisance' legislation. Initially, the nuisance provisions were contained in the Public Health Acts but more streamlined provisions were introduced into the EPA 1990. This provides definitions of 'statutory nuisances' many of which have been defined by case law. Under the EPA 1990 it is not necessary to show that an activity is prejudicial to health for it to constitute a statutory nuisance: case law has shown that an interference with personal comfort can be sufficient to establish that a nuisance exists.

Local authorities are placed under a duty to inspect their areas from time to time to identify nuisances and to investigate any complaints. Where a local authority is satisfied that a nuisance exists, or is likely to occur or reoccur, it must serve an abatement notice requiring action to be taken to remove the nuisance. Where such a notice is not complied with the authority can take action itself to remove the nuisance and recover the cost of so doing from the responsible party.

The contaminated land provisions contained in the Environment Act 1995 are modelled on the statutory nuisance provisions of the EPA 1990. Once the relevant sections of the Environment Act 1995 come into force, the statutory nuisance provisions in the EPA 1990 will cease to be applicable where the nuisance being considered is 'contaminated land'.

Environmental Protection Legislation. The environmental protection legislation is intended to prevent or control the release of harmful substances, energy, noise or vibration into the environment, thereby protecting human populations, other living organisms and the abiotic environment.

Contaminated land may have many direct or indirect consequences for environmental quality. Hazardous substances may be released slowly over time and in response to wind or water action (*e.g.* leaching from uncontained or leaking process plant, stockpiles, waste disposal areas), or through sudden events (*e.g.* failure of a fuel pipe or tank, accidents or fires); they may be released through activities designed to investigate or remediate a site, such as the rupture of a sub-surface process tank with drilling equipment, or release of vapours and dusts during excavation operations.

Responsibilities for environmental protection may arise in a wide variety of different circumstances and relate to all the main environmental media and/or to defined biological units (*e.g.* protected ecosystems or species). Some of the main legal provisions relevant to contaminated land are summarized in Table 6.

Table 5 Health and safety regulations which may be relevant to contaminated land

Management of Health & Safety at Work Regulations 1992	Make explicit many of the general duties contained in the Health & Safety at Work, *etc.*, Act 1974. They are designed to encourage a more systematic and better organized approach to health and safety matters, and impose comprehensive duties on employers to undertake suitable and sufficient assessments of health and safety risks
Control of Substances Hazardous to Health Regulations 1994	Contain many of the obligations set out in the MHSW Regulations but apply specifically to potential exposure to hazardous substances. Principal duty of the employer is to prevent the exposure of employees to hazardous substances where reasonably practicable and, in other cases, to control exposures adequately by means other than the use of protective clothing and equipment. Only where such control measures are not possible, is protective clothing, *etc.*, to be used as the control strategy
Construction (Design & Management) Regulations 1994	Designed to ensure health and safety matters are addressed throughout the lifecycle of construction projects from initial concept, through design, construction, maintenance and refurbishment, to final demolition. Places specific responsibilities for health and safety on all those involved, including clients, designers and contractors
Personal Protective Equipment at Work Regulations, 1992	Requires employers to ensure suitable personal protective equipment is available to employees where health and safety risks cannot be avoided and to carry out an assessment to ensure its suitability before the equipment is provided for use
Legislation on Packaging & Transport of Dangerous Substances	A group of regulations adopted between 1992 and 1996, including the Chemicals (Hazard Information and Packaging) Regulations (CHIP) 1993, Chemicals (Hazard Information and Packaging for Supply) Regulations 1994 and 1966 amendment, and various transport-related regulations. This aim is to improve safety performance by providing more comprehensive information, improved handling and transport arrangements, better training, *etc.*, for operatives

The Environment Act 1995

Section 57 of the Environment Act 1995 contains important new provisions on the regulation of contaminated land in England, Wales and Scotland. It inserts a new Part IIA into the Environmental Protection Act 1990 and places a duty on local authorities to inspect their areas for the purposes of identifying land which falls within a new statutory definition of contaminated land. Land formally

Table 6 Main environmental protection provisions relevant to contaminated land

Air Quality		
Clean Air Act 1993	Requires measures to be taken to prevent emissions of smoke (*e.g.* on demolition/construction projects)	
EPA 1990	Requires compliance with national air quality standards for ambient air (*e.g.* during remediation projects)	
EPA 1990	Requires authorization of the operation of specific types of processing plant (*e.g.* mobile treatment plant)	
Water Quality		
Water Resources Act 1991	<ul><li>Requires prior consent for the abstraction of water from a source of supply</li><li>Provides for the classification of water according to use</li><li>Creates an offence of causing or knowingly permitting discharges to controlled waters except in accordance with consents given under specific legal provisions</li><li>Provides powers to prevent and control pollution, and to remedy or forestall the pollution of controlled waters</li></ul>	
Water Industries Act 1991	Requires prior consent for discharges to sewers	
Waste Management		
EPA 1990	Prohibits the deposit, treatment, keeping or disposing (or knowingly doing any of the foregoing) of controlled waste in or on any land except under and in accordance with a waste management licence, and the management of controlled waste in any manner likely to cause pollution of the environment or harm to human health	
EPA 1990; Waste Management Licensing Regulations 1994; Control of Pollution (Special Wastes) Regulations 1988*	<ul><li>Defines controlled wastes</li><li>Defines 'special waste' and requires pre-notification and documentation on composition of such wastes, identity of producer and point of origin</li><li>Sets out specific requirements on the licensing of waste management activities including granting of licences, variations, revocation and suspension, surrender and supervision of waste management licences</li><li>Creates a Duty of Care on all those involved in the production, handling and disposal of controlled wastes from the point of production to the point of final disposal</li><li>Requires the registration of waste brokers</li></ul>	
Flora and fauna		
Wildlife & Countryside Act 1981; Conservation (Natural Habitats, *etc.*) Regulations 1994	<ul><li>Provides for the designation of defined areas of land with special ecological or scientific value as Sites of Special Scientific Interest (SSSI)</li><li>Affords special protection to various species of plants and animals</li></ul>	

*Note that new Regulations on Special Wastes came into force in September 1996

designated as 'contaminated land' is subject to a number of provisions intended to ensure unacceptable risks to health and the environment are properly controlled. Both local authorities and the respective Environment Agencies in England and Wales, and in Scotland, have a role to play in achieving this objective.

Contaminated land is defined in the legislation as: 'Any land which appears to the local authority in whose area it is situated to be in such a condition, by reason of substances in, on or under the land, that:

- significant harm is being caused or there is a significant possibility of such harm being caused, or
- pollution of controlled waters is being, or is likely to be, caused'

The definition makes clear that not all land containing measurable concentrations of contaminants is expected to fall within the scope of the legislation. The type and degree of harm to be taken into account, what is to be regarded as 'significant possibility' and how the remaining provisions of the legislation are to be discharged are to be set out in statutory guidance to be issued by the Secretary of State. Since the statutory guidance will contain much of the detailed advice to regulators and others on the intended application of the legislation, it forms an extremely important part of the overall regulatory regime. Under a negative resolution procedure, the statutory guidance must be set before Parliament for 40 days before it can be adopted.

At the time of writing, the statutory guidance is being prepared for public consultation. The House of Commons Select Committee on the Environment has also announced that it will conduct a short examination of the draft guidance.

The main provisions and scope of Part IIA, which generally reflect the policy blueprint set out in *Framework for Contaminated Land*, are set out in Table 7.

4 International Collaboration

There are several international networks engaged in the exchange of information and experience on contaminated land. The UK Government is currently represented on several international forums covering a range of policy and technical issues. Other organizations, such as the industry group NICOLE, are involved in similar networks aimed at international co-operation on scientific and technical issues.

Ad Hoc Working Group

The need for the Ad Hoc Working Group on contaminated land was agreed in Washington, DC, in 1993.[35] The objective of the group is: 'To provide a forum, open to any country, in which the issues and problems of contaminated land and groundwater can be discussed at an international level and information can be freely exchanged to the benefit of all participants'.

The first technical discussions of the group, hosted by Austria and with a UK

[35] Department of the Environment, *Ad Hoc International Working Group on Contaminated Land*, Report of meeting in Vienna, November 1993, DoE, London.

Table 7 Main provisions of Part IIA of the EPA 1990	Inspection and identification	Local authorities are placed under a statutory duty to inspect their areas from time to time to identify land which falls within the statutory definition. In performing this duty, local authorities are to act in accordance with statutory guidance
	Consideration of possible status as a 'Special Site'	Types of land to be considered for 'special site' status to be set out in regulations but are likely to be sites posing particular problems which require the specialist expertise available within the Environment Agency for their regulation
	Consideration of role of other statutory provisions	Whether contamination should be addressed under other statutory provisions, *e.g.* IPC, waste management or Water Resources Act 1991
	Consideration of the need for urgent action	Where urgent action is required to deal with contamination enforcing authorities have powers to act immediately (thus by-passing consultation with the appropriate persons) and to recover the costs of taking the action
	Preliminary identification of appropriate persons	Appropriate persons are those who caused or knowingly permitted the substances to be in, on or under the land (*i.e.* the original polluter) or, where no such person can be found, the owner or occupier of the land. Detailed advice on identifying the appropriate persons, and apportioning responsibility where more than one appropriate person exists, is given in the statutory guidance
	Consultation on what is to be done by way of remediation	Before a remediation notice is served, the enforcing authority must consult with those on whom a remediation notice is to be served on what is to be done by way of remediation. A minimum three month consultation period is allowed. The authority is not permitted to specify anything by way of remediation which is unreasonable given the costs involved and the nature and extent of the harm or pollution of controlled waters
	Restrictions on the service of a notice	The enforcing authority is not permitted to serve a remediation notice where: ● it is satisfied, taking all the circumstances of the case into account, that nothing by way of remediation can be specified in a remediation notice ● it is satisfied that voluntary action will be taken by the appropriate person(s) ● the appropriate person is the authority itself ● the authority has other powers to take action itself If no notice is served, a Remediation Statement is published describing the remediation which is being carried out; if there is no reasonable remediation which could be carried out, a Remediation Declaration is published explaining the reasons for this

Table 7 Continued

Service of a remediation notice	Unless restrictions apply, the enforcing authority is required to serve a remediation notice on the person(s) who have been identified as the appropriate person(s). The notice must set out what is required to be done by way of remediation and the relative liabilities of the different parties. The service of a remediation notice must be recorded on a public register
Appeals against the service of a notice	Appeals against the service of a notice can be made within a period of 21 days. Regulations are to be issued setting out the grounds for appeal; these are likely to include that the recipient of the notice is not an appropriate person or there has been a failure to follow the statutory guidance
Prosecution for failure to comply with a remediation notice	Penalties for failing to comply with a remediation notice amount to a maximum of £20 000 for industrial, trade or business premises, with a daily fine of £2000 for each day the notice is not complied with. There are two defences: ● reasonable excuse ● the refusal or inability of other liable persons to bear their proportion of the costs
Action by the enforcing authority itself	Circumstances in which the enforcing authority itself can take action include situations where: ● there is a need to take emergency action ● a written agreement exists between the enforcing authority and the appropriate person(s) that the former should undertake the necessary works at the expense of the latter ● there has been a failure to comply with a remediation notice ● action against owners/occupiers is precluded in relation to pollution of controlled waters ● requiring the appropriate person to take action could cause hardship ● after reasonable enquiry, no appropriate person can be found
Cost recovery	Where appropriate person(s) exist, the enforcing authority has powers to recover costs although it must take hardship, and any guidance issued by the Secretary of State, into account. In England and Wales (but not Scotland), interest on outstanding costs may be imposed and the authority may recover costs by way of a charge on the land
Public registers	The Act requires enforcing authorities to keep public registers of relevant matters including remediation notices, notifications of Special Sites, and information on remediation action by those who have carried out the action
Overview and reporting	Local authorities are required to provide the respective Environment Agencies with information that will enable them to satisfy their obligation to prepare overviews of the incidence and treatment of contaminated land at a national level

secretariat, took place in Vienna in November 1993. Representatives from 17 countries and five international bodies were present and a wide range of issues discussed, including: different policy approaches; risk assessment; guideline values and quality standards; economics and land use; strategic land use considerations; remedial measures; research and information transfer; research, technology and development.

The second meeting of the Ad Hoc Working Group was held in May 1995 in Nottingham. This was hosted by the UK and the Netherlands provided the secretariat. The meeting in Nottingham decided that:

- a further meeting would be held in Spring 1997, with the Netherlands acting as host and Denmark providing the secretariat
- informal international working groups would prepare papers on third party financing, registration of contaminated land, risk management approaches and technology choices, for discussion at the 1997 meeting
- the information collected by the Group on such matters as members' policy and legislation, assessment and prioritization procedures and guideline values/soil standards, would be kept up-to-date

Common Forum on Contaminated Land

The Common Forum on Contaminated Land consists of representatives of the Member States of the European Union (EU). Its role is to:

- formulate a common view of the work that the EU and its institutions should undertake on contaminated land
- make recommendations to the European Environment Agency and the Commission on the technical issues involved
- provide a link between the EU and the wider world on contaminated land matters

CARACAS

The Concerted Action on Risk Assessment for Contaminated Sites (CARACAS) is funded by the European Commission (EC) under the Environment and Climate Programme 1994–1998.[36] It involves all 15 Member States of the EU, together with Norway, Switzerland, the United States and Australia. Representatives are drawn from the national environment ministries and scientific institutions.

The objectives of CARACAS are to:

- identify, compile, assess and review all relevant Research, Technology and Development (RTD) projects and scientific approaches for risk assessment developed in the Member States of the EU

---

[36] H. Kasamas, *CARACAS, Concerted Action on Risk Assessment for Contaminated Sites*, report to the NATO/CCMS Pilot Study (Phase II) on the evaluation of demonstrated and emerging technologies for the treatment and clean-up of contaminated land and groundwater, Adelaide, February 1996.

- propose scientific priorities for future programmes and projects of the European Commission and EU Member States
- elaborate guidelines and recommendations for assessing the risks associated with contaminated sites

The output from CARACAS is expected to influence future RTD priorities for Directorate General XII of the EC, and promote international collaborative research effort, with the aim of developing a consensus approach to risk assessment.

The first meeting of the CARACAS Co-ordination Group was in Belgium in March 1996, at which seven topic groups were identified covering human toxicology, ecological risk assessment, fate and transport of contaminants, site investigation and analysis, models, screening and guideline values, and risk assessment methodologies.

NATO/CCMS Pilot Study

The NATO Committee for Challenges to Modern Society (CCMS) Pilot Study has progressed through a number of phases since the initial study first addressed the issue of contaminated land in 1984/85.[37] The next phase of the Study ran between 1986 and 1991 with an extended remit to evaluate demonstrated technologies for the remediation of contaminated soil and groundwater. The scope of the Study has been extended even further in the current phase (1992–1996) to include the evaluation of both emerging and demonstrated remedial technologies.

Technology projects nominated by member countries are accepted into the Pilot Study by country representatives on a consensus vote. Projects within the study are evaluated during technical meetings of Pilot Study participants with the aim of establishing key technical strengths and weaknesses, potential for application across national boundaries, economic characteristics, *etc.* Both interim and final reports of the Study findings are published. Pilot Study Fellows working on topics of relevance to the Study, and invited speakers, make additional contributions to the Pilot Study meetings and reports.

Thirteen different countries are actively involved in the current phase of the Pilot Study. The final report is expected to be published in 1997.

5 Conclusions

Contaminated land is a complex policy area which has attracted a good deal of public attention in the United Kingdom over the past few years. Account has been taken of the legal, technical and economic issues involved. Policy is the product of the balance struck between these different influences. Each country has its own institutions and cultural traditions which sometimes lead to the adoption of particular solutions to the problems of contaminated land but, at the

---

[37] NATO/CCMS, *Contaminated Land: Reclamation and Treatment*, NATO CCMS vol. 8, Plenum Press, New York, 1985; note that details on the NATO/CCMS Pilot Study are available from the Centre for Research into the Built Environment, Nottingham Trent University, Burton Street, Nottingham.

same time, there are many common issues which make international collaboration and the exchange of information and ideas very worthwhile.

6   Acknowledgement

The Director General of CIRIA and the Department of the Environment gave permission to the authors to use material prepared under a research project jointly funded by the Department and by the construction industry. The views expressed in this chapter are those of the authors, and are not necessarily those of CIRIA or the Department.

Remediation Methods for Contaminated Sites

PETER A. WOOD

1 Introduction

Contaminated land in the UK arises largely as a result of past industrial processes which have left a legacy of many hazardous substances in the ground, including heavy metals, organic compounds, oils and tars, and soluble salts. These substances can represent an actual or potential threat to the environment, to man, or occasionally to other targets. Although much of the contaminated land is the result of past activities associated with the industrial revolution, many contaminative uses are still in operation today. As a result, an approach for the adequate management of contaminated land should be twofold: remedial works for land that is already contaminated, and the implementation of appropriate management procedures to prevent or minimize future contamination. This chapter is concerned with remediation methods that are designed to achieve the former.

2 The Need for Remediation

The primary aims of any remediation are as follows:

- reduction of actual or potential environmental threat
- reduction of potential risks so that unacceptable risks are reduced to acceptable levels

Consequently, the need for any remediation will depend on the degree of any actual or potential environmental threat or the level of any risk. Aspects of risk will in turn depend partly on the expected end-use of the site following remediation, as different at-risk targets can be associated with different end-uses.

Remediation may be required for the treatment of one or more of the following:

- contamination source
- contaminated soil
- contaminated debris
- contaminated groundwater
- contaminated soil atmosphere

3 Definitions of Types of Remediation

Remediation of a contaminated site is achieved by one or more of the following objectives:

- removal or destruction of the contaminants
- modification of the contaminants to a less toxic, mobile or reactive form
- isolation of the contaminant from the target by interrupting the pathway of exposure

Although a wide range of remediation methods are available to achieve these objectives, two broad approaches can be distinguished:

- engineering approaches; these are primarily the traditional methods of excavation and disposal to landfill or the use of appropriate containment systems
- process-based techniques; these include physical, biological, chemical, stabilization/solidification and thermal processes

Landfill involves the three stages of soil excavation, transport and burial at the landfill site. Contaminants in the soil are not necessarily removed, stabilized or destroyed on site and are ultimately transferred to another site. Landfills are designed to ensure that contaminants are either isolated from the environment or subjected to attenuation processes so that they no longer cause harm to the environment.

Containment measures are those which are designed to prevent or limit the migration of contaminants, that may be either left in place or confined to a specific storage area, to the wider environment. Approaches include hydraulic measures, capping, use of break layers and low permeability barriers, amongst others.

Physical processes used in soil treatment are used to remove contaminants from the soil matrix, concentrating them in process residues that require further treatment or safe disposal. Contaminants in the concentrated fractions may subsequently be destroyed or recovered by some other process (*e.g.* chemical or thermal) or may be disposed of to landfill.

Biological processes of soil treatment depend on the natural physiological processes of micro-organisms, such as bacteria and fungi, to transform, destroy, fix or mobilize contaminants. Biological processes can also be used to fix and accumulate contaminants in a harvestable biomass; such a process may use higher plants.

Chemical processes in soil treatment systems are used to destroy, fix or neutralize hazardous compounds. Many processes in other categories may use chemical processes for the treatment of effluents and gaseous emissions.

Stabilization/solidification processes involve solidifying contaminated materials, converting contaminants into less mobile chemical forms and/or binding them within an insoluble matrix, presenting a minimal surface area to leaching agents.

It is when the process results in chemical fixation of contaminating substances that the term stabilization can be applied.

Thermal processes use heat to remove or destroy contaminants by incineration, gasification, desorption, volatilization, pyrolysis or some combination of these.

Most remediation technologies aim for rapid remediation of a site, making intensive use of energy and other resources. Recently, however, interest has grown in using lower input technologies which take longer to become effective but have lower cost and management requirements. These technologies have been called extensive.[1] Such extensive technologies can make use of both engineering and process-based approaches.

Many of the process-based approaches can be applied either *in situ* or *ex situ*. Each approach has specific advantages and disadvantages. Additionally, many of the methods, both *in situ* and *ex situ*, are at various stages of development and may not be available commercially. The following sections discuss the various remediation options, with emphasis on those approaches that are technologically advanced or commercially available.

4 Removal to Landfill

Contaminated sites are frequently remediated by excavation of the contaminated material and subsequent disposal of this to a controlled landfill. The approach represents a rapid method of dealing with a contaminated site, but it has been criticized as it represents only a transfer of the contaminated material from one location to another rather than a final solution. Relatively low landfill disposal costs are the major incentive for this type of disposal, although some recent increases in the cost of disposal of hazardous material to landfill may result in some reduction in this approach. The major advantages and disadvantages of removal to landfill are outlined in Table 1.

Containment and Attenuation

Contaminated material disposed of to landfill must be prevented from causing any further environmental damage. The principal approaches that contribute towards prevention are:[2]

- containment
- attenuation

Containment measures at a landfill are designed to isolate the disposed material from the environment such that any liquid or gaseous interchange is minimized or controlled. This isolation may be achieved by a range of containment techniques, including lining, capping, cover systems and, on occasion, vertical barrier systems (see Section 5). Some of these containment techniques may also be used for the on-site disposal of excavated contaminated

---

[1] P. Bardos and H. van Veen, *Land Contam. Reclam.*, 1996, **4**, 19.

[2] R. Armishaw, P. Bardos, R. Dunn, J. Hill, M. Pearl, T. Rampling and P. Wood, *Review of Innovative Contaminated Soil Clean-up Processes*, LR 819 (MR), Warren Spring Laboratory, Stevenage, 1992.

Table 1 Advantages and disadvantages of disposal to landfill[2,3]

Advantages	Disadvantages
Wide range of contaminants and materials can be disposed of	Contaminants are only removed to different location
Contaminants are completely removed from a site	Quality of containment may be difficult to ensure
Can be relatively low cost	Long-term integrity not well understood
Permits rapid remediation of site	Long-term monitoring required
Asociated excavation offers advantages of improving ground conditions or recycling selected materials	Leachate and gas collection systems may be required

material, which obviates the need for transport to a landfill. Control of any leachate or gaseous products may also be achieved by leachate or gas collection systems. The effective design and installation of a containment system requires extensive geological and hydrological investigation, modelling and monitoring. Although low permeability is a necessary characteristic of containment materials, complete impermeability is rarely attained in practice. However, any materials used for containment may also act as a substrate for attenuation mechanisms. A further degree of containment can be achieved if the contaminated material is subjected to solidification/stabilization techniques.

Attenuation occurs as a result of various mechanisms operating in the landfill which serve to minimize the movement and/or reduce the toxicity of contaminants.[4] Attenuation mechanisms can be physical, chemical or biological. Physical methods include the adsorption and absorption of contaminants, filtration, dilution and dispersion. Chemical methods of attenuation include acid–base interactions, oxidation, reduction, precipitation and ion exchange. Biological methods of attenuation include aerobic and anaerobic microbial degradation. Most of these mechanisms require the presence of organic material within the landfill and so co-disposal operations are favourable as these include the disposal of decomposing municipal waste.

5 Containment

The concept of containment as a method for dealing with contaminated ground is based on the use of low permeability barriers to isolate the contaminated material, or any associated leachate or gaseous products, from the environment. The barriers can be constructed from natural or synthetic materials, or a combination of both, and can be placed over, under or around a contaminated area or pollution source. The technique can be used to isolate existing hazards such as an inherited contaminated site, to prevent the spread of contaminants from a disposal site such as landfill, or to isolate specially designed mono-disposal

---

[3] CIRIA, *Remedial Treatment for Contaminated Land*, Special Publications 101–112, CIRIA, London, 1995, vols. 1–12.

[4] DoE, *Landfilling Wastes: A Technical Memorandum for the Disposal of Wastes on Landfill Sites*, Waste Management paper No 26, HMSO, London, 1986.

sites for contaminated soil. Landfills receiving controlled waste will incorporate containment systems.

The remediation of many contaminated sites has been achieved by covering the surface with clean material incorporating a low permeability layer. Whereas this may reduce infiltration and form a physical barrier to the contamination, it may not necessarily control adequately the movement of contaminants. In order to provide adequate control it may be necessary to use such cover systems in conjunction with vertical and horizontal in-ground barriers or cut-offs to achieve partial or total isolation of the site. This isolation, in its extreme, can involve the complete enclosure of a site within an impermeable barrier to prevent the ingress or egress of water and contaminants in all directions. The following types of containment will be examined:

- cover systems
- in-ground barriers

Cover Systems

A cover system consists of a single layer, or succession of layers, of selected non-contaminated material that covers the area of contamination. The purpose of a cover system is to provide specific physical and chemical properties such that one or more of the following is achieved:[2,3,5]

- prevention of exposure of at-risk targets to potentially harmful substances; these targets might include humans, vegetation, animals, buildings and the environment
- sustained growth of vegetation
- control of infiltration to, and subsequently from, the site
- any geotechnical requirements satisfied

Table 2 details the requirements of a cover system and indicates the purpose of the various components that might be incorporated.

The choice of materials suitable for use as a cover system is wide and would depend on the physical properties required by any particular component layer. Possible materials include:

- natural clays, sub-soils and soils
- amended soils incorporating materials like pulverized fuel ash, lime and sludges
- waste materials such as fly ash, slags, dredgings, sewage sludges
- synthetic membranes and geotextiles
- concrete, asphalt, *etc.*

In practice, economic factors, in addition to physical properties, will influence the material selected and a compromise between ideal and locally available materials might be necessary. In this way, a cover system can represent an effective treatment procedure at reasonable cost.

[5] T. Cairney, in *Reclaiming Contaminated Land*, ed. T. Cairney, Blackie, London, pp. 144–169.

Table 2 Requirements and components of a cover system[2,5,6]

Requirements	Component
Minimize toxicity in upper layers of soil	All component layers: to provide sufficient thickness of clean materials
Limit surface water percolation and minimize leachate	Top soil: to support vegetation
Control vertical gas movement	Sub-soil: to support vegetation
Prevent capillary movement of contaminants	Low permeability barrier layer: to prevent passage of water, gas or VOCs
Prevent soil erosion and dust generation	Buffer layer: to protect barrier layer
Support vegetation	Drainage layer/system: control drainage
Inhibit root penetration into contaminated layers	Gas control layer/system: gas control
Improve structural properties to facilitate road or building construction	Filter layer: control the movement of particles
Improve aesthetics	Break layer: to prevent capillary movement of contaminants

Cover systems have been used widely within the UK for the reclamation of many types of contaminated sites such as gasworks, tarworks, waste disposal sites, metal mining areas, chemical works and tanneries. Many of these have been successful but, because the contaminants still remain on site, a cover system is not necessarily an effective long-term solution in those cases where contaminants are mobile. Cover systems do not necessarily control groundwater movements, gaseous emissions and odours and, following completion, long-term monitoring may be necessary.

Quality assurance is of vital importance during the installation of a cover system, whatever its design and whatever the type of materials used in construction. It is essential that the desired properties of individual component layers are maintained over the entire area and depth. Integrity must not be compromised by inappropriate or poor quality materials. Careless or inappropriate use of heavy construction equipment can damage previously completed layers, particularly when synthetics are used. The major advantages and disadvantages of a cover system are identified in Table 3.

In-ground Barriers

In-ground barriers can be used to isolate, usually by physical means, a contaminated mass of ground from the surrounding environment or other targets. Low permeability material may be introduced around or under the contaminated site, or methods incorporating some sort of physical, biological or chemical control of contaminant migration can be used. In-ground barriers can be placed around, above and below a contaminated mass to achieve complete isolation, a method that has been termed macroencapsulation. In practice,

[6] S. Barber, P. Bardos, H. van Ommen, J. Staps, P. Wood and I. Martin, *Contaminated Land Treatment: Technology Catalogue*, European Commission, DG XI, Brussels, 1994.

Table 3 Advantages and disadvantages of cover systems[2,3,5]

Advantages	Disadvantages
Wide range of contaminants and materials can be isolated	Contaminants are not removed from the site
Can be relatively low cost	Possibility for failure over long term
Permits rapid remediation of the site	Leachate and gas collection systems may be required
Suitable for use on large sites where large volumes of material might be involved	Groundwater movements not necessarily controlled
	Long-term monitoring may be necessary
	Possible long-term restrictions on use of site
	Need to consider contingency liability and insurance implications

however, it is often difficult to ensure the continuity of any barrier, particularly if underground obstructions are present.

Vertical barriers can be constructed by various methods:[2,3]

- excavation: excavation barriers require the removal of material to form a trench followed by the backfill of the barrier material; the types of material that can be used include clays, clay–bentonite mixtures and concrete diaphragm walls
- displacement: displacement barriers do not require any excavation prior to excavation; instead, the barrier material or form work is inserted by force directly into the ground; particular types of displacement barriers include steel sheet piling, vibrated beam slurry walls and membrane walls
- injection: barrier material may be injected under pressure, often incorporating a mixing process to mix the injected material with soil components *in situ*; injection methods include chemical grouting, jet grouting, jet mixing and auger mixing

The effectiveness and applicability of barrier methods vary according to the types and nature of contaminants present, the physical conditions of the site and the design life of the barrier. The long-term integrity of barrier materials may, in some cases where specific contaminants are present, not be known. Also, at present there is only limited experience with the installation of some barrier types. For example, horizontal in-ground barriers have only been installed in a few instances. Particular problems with their installation include ensuring the continuity of the barrier under the entire contaminated area. The main advantages and disadvantages of using in-ground barriers are listed in Table 4.

6 Physical Processes

Physical processes separate contaminants from uncontaminated material by exploiting differences in their physical properties (*e.g.* density, particle size,

Table 4 Advantages and disadvantages of in-ground barriers[2,3,6]

Advantages	Disadvantages
Wide range of contaminants and materials can be isolated Can be economic where large volumes of contaminated material are involved Can be used to control solid, liquid and gaseous contaminants	Contaminants remain on site Installation difficulties if underground obstructions present Difficulties in ensuring barrier continuity Limited understanding of long-term integrity Long-term monitoring may be necessary Possible long-term restrictions on use of site Need to consider contingency liability and insurance implications

volatility) by applying some external force (*e.g.* abrasion) or by altering some physical characteristic to enable separation to occur (*e.g.* flotation). Depending on the nature and distribution of the contamination within the soil, physical processes may result in the segregation of differentially contaminated fractions (for example, a relatively uncontaminated material and a contaminant concentrate based on a size separation) or separation of the contaminants (for example, oil or metal particles) from the soil particles. Table 5 summarizes the main advantages and disadvantages of physical processes.

The range of physical processes includes a diverse variety of methods that include both *in situ* and *ex situ* approaches. This variation has been classified into two main groups:[6]

- washing and sorting treatments
- extraction treatments

Washing and Sorting Treatments

Washing and sorting treatments are commonly referred to as soil separation and washing. The main aim of the processes is to concentrate the contaminants into a relatively small volume so that the costs associated with disposal and further treatment are related only to the reduced volume of process residues.

Washing and sorting falls into two main categories:

- separation from the soil of those particles containing the contaminants by mineral processing techniques, exploiting differences in the properties of individual soil particles; the volume of contaminated material requiring further treatment or disposal is thereby reduced
- transfer of the contaminant from particle surfaces into an aqueous phase by leaching using liquid extractants or steam; the contaminant-rich liquor can then be treated as waste water

Table 5 Advantages and disadvantages of physical processes[2,3,6]

Advantages	Disadvantages
Some methods are already or are becoming established	Secondary waste streams may require treatment or disposal
Potential to reduce volume of material requiring disposal or expensive treatment	Soils with high clay or peat content may be difficult to treat
Wide range of contaminants treatable	Use of some solvents will have health and safety implications
Wide range of materials treatable	Quality assurance measures needed, especially for *in situ* methods
Some *in situ* methods require only little site disruption	Approval by regulatory authority may be needed
Mobile plant available for some methods	

Washing and sorting treatments have been used in several countries, particularly Germany and the Netherlands, for the treatment of a range of soils and contaminants. However, their use in the UK is currently very restricted.

Extraction Treatments

Extraction treatments involve processes that remove the contaminants from soils by involving a mobilizing and/or releasing process to remove the contaminant from the soil matrix. Three main categories of extraction treatments are:

- soil vapour extraction: an *in situ* process where a vacuum is applied through extraction wells to create a pressure gradient that induces gas phase volatile contaminants to flow through the soil to the extraction wells, where they then become removed from the soil
- electroremediation: an *in situ* process where an electrical current is passed through an array of electrodes that is embedded in the soil; when the current is applied, movement of contaminants in the pore water towards the electrodes is induced by electrolysis, electro-osmosis and electrophoresis; the electrodes have porous housings into which purging solutions are pumped to remove the contaminants and bring them to the surface; the purging solutions are then pumped to a water treatment plant for contaminant removal
- soil flushing and chemical extraction: processes that use chemical reagents, solutions or steam to mobilize and extract contaminants from soils; mobilization refers to the release of dissolved contaminant ions from sorbed or precipitated forms in soils and may form part of both *in situ* soil flushing and *ex situ* chemical extraction treatments

7 Biological Processes

The objective of a biological remediation process is the degradation of contaminants to harmless intermediates and end products. The ultimate aim is the complete mineralization of contaminants to carbon dioxide, water and simple inorganic compounds. A large number of organic contaminants can be degraded by micro-organisms and most biological treatments attempt to optimize conditions for degradation by the naturally occurring indigenous microbial population. Achieving these optimum conditions may require control of temperature, oxygen or methane concentrations, moisture content and nutrients.

Biological degradation processes can be either *in situ* or *ex situ* and either aerobic or anaerobic. Biological treatments have considerable scope for integration with other remediation processes and are applicable to both contaminated soil and groundwater. An advantage of the simpler biological treatments is their potential to be cost effective, although long treatment times may be necessary. The presence of certain contaminants such as pesticides or heavy metals, however, may inhibit the effectiveness of a biological treatment. An additional problem is the possible creation of more hazardous intermediate products.

A range of biodegradation processes are available and methods of classification vary. Main groups include:

- *in situ* processes: involving the injection of air or water to convey oxygen and possibly nutrients into the underground contaminated mass
- dynamic *ex situ* processes: following extraction dynamic processes, in addition to the control of water, nutrients, *etc.*, are applied to mix the soil and encourage rapid degradation; processes include landfarming, windrow turning and bioreactors
- static *ex situ* processes: following extraction, the material is left undisturbed for the duration of the treatment, possibly under a liner; the condition of the material can be monitored and possible addition made of water, nutrients and air; the main static process is soil heap bioremediation, sometimes referred to as biopiles and composting
- bioaccumulation: bioaccumulation has been identified as a method for the possible treatment of soils contaminated with heavy metals; hyperaccumulator plants, capable of mobilizing and recovering metals, are grown on the contaminated area and then harvested to remove the metals from the site

Table 6 summarizes the main advantages and disadvantages of biological processes.

8 Chemical Processes

Chemical treatment processes for the remediation of contaminated soil are designed either to destroy contaminants or to convert them to less environmentally hazardous forms. Chemical reagents are added to the soil to bring about the appropriate reaction. In general, excess reagents may need to be added to ensure that the treatment is complete. This in turn may result in excessive quantities of unreacted reagents remaining in the soil following treatment. Heat and mixing may also be necessary to support the chemical reaction. Chemical processes can also concentrate contaminants in a manner similar to physical processes.

A range of chemical remediation processes are at various stages of development, both for *in situ* and *ex situ* applications. Many of these are based on the treatment

Table 6 Advantages and disadvantages of biological processes[2,3,6]

Advantages	Disadvantages
Applicable to both contaminated soil and groundwater	High cost of complex processes
Potential for integration with other processes	May require long process times
Simple processes can be cost effective	Possible formation of hazardous intermediate products
High contaminant specificity is possible	Presence of some contaminants may inhibit degradation
	Most inorganic contaminants may not be treatable
	Some complex organic contaminants may not be treatable

of waste water or other hazardous waste. However, the range of processes that have been widely used at full scale is restricted. Major types include:

- oxidation–reduction
- dechlorination
- extraction
- hydrolysis
- pH adjustment

The major advantages and disadvantages of chemical remediation methods are summarized in Table 7.

Oxidation–Reduction

An oxidation–reduction (redox) reaction is a chemical reaction in which electrons are transferred completely from one chemical species to another. The chemical that loses electrons is oxidized while the one that gains electrons is reduced. Redox reactions can be applied to soil remediation to achieve a reduction of toxicity or a reduction in solubility.

Oxidation and reduction processes can treat a range of contaminants, including organic compounds and heavy metals. Oxidizing agents that can be used include oxygen, ozone, ozone and ultraviolet light, hydrogen peroxide, chlorine gas and various chlorine compounds. Reducing agents that can be used include aluminium, sodium and zinc metals, alkaline polyethylene glycols and some specific iron compounds.

Dechlorination

Chemical dechlorination processes use reducing reagents to remove chlorine atoms from hazardous chlorinated molecules to leave less hazardous compounds. Dechlorination can be used to treat soils and waste contaminated with volatile halogenated hydrocarbons, polychlorinated biphenyls and organochlorine pesticides.

Table 7 Advantages and disadvantages of chemical processes[2,3]

Advantages	Disadvantages
Applicable to wide range of matrix types if good mixing/contact achieved	Effectiveness requires good mixing/contact which may be difficult with some soils
High degree of chemical specificity possible	Unreacted chemical reagents may remain in the soil
	Any intermediate or by-products may be hazardous
	Pre-processing may be needed to remove debris, for size reduction or to form slurry
	Effective environmental control of *in situ* methods difficult

Extraction

Extraction techniques that can be used for the treatment of contaminated soil include organic solvent extraction, supercritical extraction and metal extraction using acids. The methods are applicable to soils, waste, sludges and liquids. Following extraction of the contaminant, the extraction liquid containing the contaminant has to be collected for treatment.

Hydrolysis

Hydrolysis refers to the displacement of a functional group on an organic molecule with a hydroxy group derived from water. A restricted range of organic contaminants are potentially treatable by hydrolysis, although hydrolysis products may be as hazardous as, or even more hazardous than, the original contaminant.

pH Adjustment

pH adjustment refers to the application of weakly acidic or basic materials to the soil or groundwater to adjust the pH to acceptable levels. A common example is the addition of lime to neutralize acidic agricultural soils. Neutralization can also be used to affect the mobility or availability of contaminants such as metals by enhancing their precipitation as hydroxides.

9 Stabilization/Solidification

Stabilization/solidification methods operate by solidifying contaminated material, converting contaminants into a less mobile chemical form and/or by binding them within an insoluble matrix offering low leaching characteristics. These processes can be used to treat soils, wastes, sludges and even liquids, and a variety of contaminants types. However, the treatment of organic contaminants is generally more difficult and more expensive. Many of the reagents used for stabilization/solidification are proprietary products. An added benefit is the improved handling and geotechnical properties of the treated product that might result compared with the original contaminated material.

Stabilization/solidification processes have been applied both *in situ* and *ex situ*, the latter being both on and off site. With an *ex situ* approach it may be necessary to landfill the stabilized product if an alternative use or disposal option is not possible. A disadvantage here is that the volume of the stabilized product can be considerably greater than the original contaminated material because of the quantities of stabilization materials that have been added.

Solidification/stabilization processes can be classified according to the type of material used as the binder. The more frequently applied methods use:

- Portland cement
- pozzolanic materials such as fly ash
- lime
- silicates
- clays (often used in conjunction with other materials)
- polymers (often used in conjunction with other materials)
- other proprietary additives

Table 8 summarizes the advantages and disadvantages of solidification/stabilization.

10   Thermal Processes

A wide range of thermal processes are at various stages of development,[2,6] although the number of technologies that are commercially available is considerably more restricted. Techniques under development and commercially available can be either *in situ* or *ex situ*. Three *ex situ* techniques will be outlined that operate in different temperature regimes:

- thermal desorption
- incineration
- vitrification

The advantages and disadvantages of thermal processes are summarized in Table 9.

Thermal Desorption

Thermal desorption involves excavation of the contaminated soil followed by heating to temperatures in the region of 600 °C. At these temperatures the volatile contaminants are evaporated and subsequently removed from the exhaust gases by condensation, scrubbing, filtration or destruction at higher temperatures.[3] Following treatment it may be possible to re-use the soil, depending on the temperatures used and the concentration of any residual contamination. Thermal desorption has its primary use in the treatment of organic contamination, although it has also been used for the treatment of mercury contaminated soils.

Table 8 Advantages and disadvantages of solidification/ stabilization[2,3,6]

Advantages	Disadvantages
Applicable to inorganic and organic contaminants, although organics are less proven	Contaminants contained rather than destroyed or detoxified
Applicable to soils, sludges and liquids	Increase in volume of material following treatment
Improved handling and geotechnical properties possible	Some processes produce heat which can cause gaseous emissions
Rapid treatment possible	Quality assurance measures needed, especially for *in situ* methods
Ex situ methods relatively easy to apply	Uncertainties over long-term performance, especially with organic contaminants
	Long-term monitoring required

Table 9 Advantages and disadvantages of thermal processes[2,3,6]

Advantages	Disadvantages
Potential for complete destruction of contaminants	High cost of some methods due to high energy requirements
Applicable to a wide range of soil types although may be handling problems	Soil may be destroyed by high temperatures
Established technologies with some mobile plant for selected process types	Heavy metals contaminants may not be removed and may become concentrated in ash
Possible re-use of soil if process temperature is not excessive	Potential for generation of harmful combustion products
	Control of atmospheric emissions required including condensing of volatile metals
	Approval by regulatory authority may be required

Incineration

Incineration involves the heating (either directly or indirectly) of excavated soil to temperatures between 880 and 1200 °C to destroy or detoxify contaminants. Incineration can also be used for the treatment of contaminated liquids and sludges. Incineration results in the destruction of the soil texture and removes all natural humic components. The residues may also have high heavy-metal contents. Exhaust gases need to be treated to remove particulates and any harmful combustion products. A range of methods of incineration are available, although the use of rotary kilns is probably the most widespread. Costs of treatment are heavily dependent on the water content of the material being treated and any calorific value that the material may have.

Vitrification

Vitrification involves the heating of excavated soil to temperatures in the region of 1000–1700 °C. At these temperatures, vitrification of the soil occurs, forming a monolithic solid glassy product. Contaminants will either be destroyed or trapped in the glassy product. The technology works by melting the aluminosilicate minerals in the soil which, on cooling, solidify to form the glass. In soils or waste where there are insufficient aluminosilicates, these can be added in the form of glass or clay. The product from vitrification may have very low leaching characteristics. Exhaust gases require treatment for the removal of any volatile metals or hazardous combustion products. Vitrification is an expensive process and likely to be restricted in use for particularly hazardous contaminants that are not readily treated by other methods.

11 Process Integration

The process-based technologies available to remediate contaminated sites have been examined in previous sections. They can generally be grouped into five categories:

- physical treatment
- chemical treatment
- biological treatment
- thermal treatment
- treatment by stabilization and solidification

Additionally, other techniques are available to contain off-site contaminant migration, *e.g.* impermeable barrier walls, hydraulic measures and cover systems, or contaminated material may be excavated for disposal elsewhere.

Although individual unit processes may have proven track records for treating a wide variety of contaminated soils, there are site-specific limitations, often related to either complex mixtures of contaminants or to the nature of the soil constituents, that may make the successful application of the techniques ineffective or uneconomic. One approach of extending the applicability and cost effectiveness of individual techniques is to use more than one treatment on a site that the combination results in the overall successful and economic treatment of the site. Different unit processes could either be used in different parts of the site to treat different areas or materials as appropriate, or be used consecutively to treat and re-treat the same area or material. Alternatively, methods can be used to reduce a complex contamination problem into several simpler problems. This general concept of combining technologies is known variously as process integration, treatment trains, bundled technologies or treatment combination.

In its broadest sense, process integration can also be extended to include the combination of processes to treat groundwater as well as soil contamination. This has particular relevance to the combination of *in situ* processes.

Process integration aims to minimize costs for treatment of complex contamination problems by maximizing the use of low cost, bulk treatments and reserving the higher cost methods for a more limited volume of material. The main objective of process integration is to enhance soil treatment by extending the potential application of individual methods beyond that where they would normally be used as a single, stand-alone treatment.[7] Thus the individual processes that might be involved in process integration can be used for:

- pretreatment: preparation of the soil/contaminant mixture for treatment by subsequent processes
- treatment: treatment of the soil, or parts of the soil, either in parallel or in succession
- combined pretreatment and treatment: some processes may be a treatment stage but at the same time may also be acting as a pretreatment stage for a subsequent treatment; alternatively, some individual processes are multi-staged and incorporate their own pre-treatment step

With some combined processes the elements of the treatment train are not easily defined, *i.e.* a clear leading treatment or need is not apparent as the technologies that form the combined processes are synergistic, each running side-by-side with attributes of one process being simultaneously exploited by another process.

The upstream treatment within a process stream aims to enable the downstream processes to be more efficient and cost effective by:

- reducing the volumes of soil requiring further treatment
- improving access to the contaminant
- changing the form of the contaminant
- changing the phase of the medium holding the contaminant
- changing the concentration of the contamination
- treating or containing mobile contaminants
- simplifying complex contamination

The upstream treatment approaches and the synergistic simultaneous concepts are discussed below, although in practical terms any particular operation may accomplish more than one of these functions.

Reducing the Volume of Soil Requiring Further Treatment

The cost effectiveness of certain treatment methods may be improved if the downstream processes treat only those soil fractions that contain the contaminants. Volume reduction methods can therefore be used to separate the relatively clean bulk from those soil fractions containing the contaminants. The contaminated fractions are then further processed.

Improving Access to Contaminants

The ease of access to contaminants by the treatment process has an important effect on the success of the treatment. An initial stage in a process integration

---

[7] M. Pearl, P. Wood and R. Swannell, *Review of Treatment Combinations for Contaminated Soil*, LR 1016, Warren Spring Laboratory, Stevenage, 1994.

approach can be undertaken to improve access to the contaminant by a subsequent treatment stage. Examples are:

- physical crushing of the material to allow access and release of the contaminants held within the material
- slurrying of a soil to encourage contact between contaminants and chemical/biological agents during a subsequent treatment
- adding chemical amendments to the soil to increase the accessibility of moisture, nutrients, air, *etc.*, during further treatment
- hydraulic or pneumatic fracturing of *in situ* soils to improve permeability and hence access to contaminants during an *in situ* process such as pump-and-treat, soil vapour extraction or bioventing
- synergistic treatment effects where the first process not only removes contaminants but also acts as an enabling stage for a second process; electroreclamation, for example, increases the permeability of the treated soil and raises the temperature so that biological processes are then enabled

Changing the Form of Contaminants

Some contaminants may not be easily treated in the form in which they occur in the soil. This may be due to problems associated with the mobility and toxicity of the contaminant, which may limit the effectiveness of a treatment. An initial stage in process integration can be undertaken to effect an appropriate change in the form of the contaminant and so improve the efficiency of any subsequent treatment. Examples are:

- changing the contaminant from a less to a more mobile form so that it can be more easily removed by a subsequent process such as leaching or soil washing
- changing the contaminant from a more to a less mobile form so that it can be more easily stabilized by a stabilization process
- changing the contaminant from a less to a more volatile form so that it can be more easily removed by, for example, a thermal or *in situ* venting process
- changing the contaminant to a form that is less recalcitrant and therefore more easily degraded by a subsequent biological process
- changing the contaminant to a form that is less toxic and therefore more amenable to degradation by a subsequent biological process.

Changing the Phase of Medium Holding the Contaminant

Some of the contaminants in a soil may be very amenable to phase transfer from the existing soil medium to a medium with a different phase. An initial stage in process integration could bring about the transfer of contaminants to a gaseous or liquid medium from which they can then be removed by a subsequent treatment process. Examples might be:

- transfer of contaminants from the solid (soil) medium to a liquid medium for subsequent removal by a waste water and effluent treatment processes

- transfer of contaminants from the solid (soil) medium to a gaseous medium for removal by a subsequent gaseous emission control process

Changing the Concentration of Contaminants

Contaminant concentrations in a soil may not be within an optimum range for a specific treatment to be either successful or cost effective. An initial stage in a process integration approach can be employed to change the concentration of contaminants in the feedstock that goes forward for subsequent treatment. Examples might be:

- increasing contaminant concentration by selective particle separation techniques in order to produce contaminant concentrates that can be cost effectively stabilized or incinerated
- decreasing the concentrations of some of the contaminants by first-stage treatment techniques so that a subsequent process can effectively treat other toxic contaminants in the residue, *e.g.* reducing the heavy metal content by a first process to enable a subsequent biological process to treat an organic contaminant-rich residue

Treating or Containing Mobile Contaminants

Sites may contain a mixture of mobile and non-mobile contaminants. Any mobile contaminants may be released into the environment during treatment if adequate control measures are not exercised. An initial stage in process integration may be to remove or contain the more mobile contaminants to prevent release of environmental damage during a subsequent treatment process which deals with the non-mobile contaminants. Examples might be:

- controlled removal of volatiles by air stripping during excavation of the soil prior to further treatment
- use of a slurry wall containment system to control groundwater contamination prior to treatment of the groundwater or the soil

Simplifying Complex Contamination

Contamination at any specific site is often complex and consists of a range of contaminants, all of which may not be treatable by any single process, even after appropriate pretreatment. When this occurs, many of the first-stage treatment methods outlined above can be employed to separate the complex contamination into a series of simpler contamination problems so that each can then be treated by an appropriate treatment method. Examples are:

- separation of complex contamination into organic and inorganic contaminants so that the organics can be biodegraded or incinerated while the inorganics can be leached or stabilized
- separation of complex contamination into volatile and non-volatile contaminants so that the volatile components can subsequently be removed

by gaseous emission control methods or *in situ* vacuum extraction processes while non-volatiles can be subjected to appropriate soil treatment methods
- separation of complex contamination into contaminants associated with a liquid medium and those associated with a solid medium; the liquid medium can then be treated by effluent treatment methods or removed by *in situ* pump-and-treat processes while the solid medium can be subjected to appropriate soil treatment techniques

Synergistic Simultaneous Processes

The application of any specific treatment process may also have consequential effects on the potential for a second. These circumstances may facilitate the employment of the second technique. For example, the bioventing soil vapour extraction processes utilizing air injection result in an increased supply of air and associated oxygen at depth within the soil. This in turn can result in enhanced biological activity and biodegradation of organic contaminants.

12   Selecting the Best Practicable Environmental Option

The remediation strategy determined for a particular site should be capable of removing any actual or potential threat to the environment and of reducing any risks associated with the contamination to an acceptable level. However, in addition to offering the necessary degree of protection the strategy should be practical and meet cost requirements.

CIRIA[3] and Harris[8] identify various factors to be considered when selecting remedial methods, while Wood and Bardos[9] discuss some of the problems associated with remediation process selection. Additionally, the Department of the Environment and National Rivers Authority have funded the production of guidelines for the evaluation and selection of remedial measures. Factors to be considered include:

- applicability
- effectiveness
- limitations
- costs
- development status
- availability
- operational requirements
- information requirements
- monitoring needs
- potential environmental impact
- health and safety needs
- post-treatment management needs

[8]   M. Harris, *Land Contam. Reclam.*, 1993, **1**, 77.

[9]   P. Wood and P. Bardos, *Constraints to Effective Contaminated Land Remediation*, LR 1003, Warren Spring Laboratory, Stevenage, 1994.

Applicability

The remediation process has to be applicable to both the contaminants and the contaminated medium, be it soil particles, groundwater or soil atmosphere. The remediation process must be able to reach the contaminants and the contaminants must be available to, and in the correct form to be treated by, the process. In practice, few processes are universally applicable and it may be necessary to use a combination of processes to achieve the desired level of remediation.

Effectiveness

Long-term effectiveness should be a major criterion in the evaluation and selection of any remediation process. The remediation process must be capable of achieving the level of treatment and risk reduction required. With engineering-based methods, effectiveness may vary through time as physical barriers may become less effective through deterioration and accidental or deliberate disturbance. With process-based remediation treatments, effectiveness could vary according to a number of factors such as the contaminant concentration, the type of soil, the feed rate to the process, the duration of treatment, *etc*. Screening and treatability studies may be necessary to aid process selection and to determine optimum process conditions. Evaluating effectiveness during remediation is likely to require monitoring and quality assurance measures.

Limitations

Various factors may constrain the use of remedial methods owing to limitations imposed by the process or the site. Process limitations may include excessive time constraints required for some biodegradation processes or the need to control atmospheric emissions from thermal treatments. Site-specific limitations may be imposed by the site location and size, or the need to maintain the site operational during the remediation.

Cost

Cost is probably the most important non-technical parameter to be considered when selecting the method of remediation for any site. However, the cost of remedial work is often difficult to predict because of the number of unknowns and variables involved, and because of difficulties in fully characterizing any site. These difficulties apply to both engineering and process-based approaches, although the latter are probably more difficult to evaluate because of a more limited track record. It is generally agreed that the costs of process-based treatments are generally substantially higher; indicative costs are provided in Table 10.

Development Status

The development of a remediation process evolves over a period of time and undergoes a transition from an emerging technology, through an innovative

Table 10 Indicative costs of selected remediation approaches[10-12]

Remediation approach	Cost range
Surface amendments	£10–40 t^{-1}
Excavation and disposal off-site	£7–50 t^{-1}
Cover system	£20–30 m^{-2}
Containment	£10–50 t^{-1}
Vertical slurry wall: shallow	£30–60 m^{-2}
Vertical slurry wall: deep	£60–120 m^{-2}
Soil washing	£50–250 t^{-1}
Physico-chemical washing	£50–170 t^{-1}
Soil flushing	£25–80 t^{-1}
In situ soil vapour extraction	£10–90 t^{-1}
In situ stabilization/solidification	£60–110 t^{-1}
In situ electrokinetic techniques	£40–120 t^{-1}
Biological treatment: bioslurry	£50–80 t^{-1}
Biological treatment: biopiles	£15–45 t^{-1}
Biological treatment: land farming	£10–100 t^{-1}
Biological treatment: windrow turning	£5–60 t^{-1}
Biological treatment: *in situ*	£5–160 t^{-1}
Biological treatment: bioventing	£15–80 t^{-1}
Thermal treatment	£40–700 t^{-1}
Incineration	£50–1200 t^{-1}
Kiln-based vitrification	£30–500 t^{-1}
Ex situ stabilization/solidification: inorganic	£20–35 t^{-1}
Ex situ stabilization/solidification: organic	£40–60 t^{-1}
Solidification: cement and pozzolan based	£20–170 t^{-1}
Solidification: lime based	£20–40 t^{-1}
Solvent extraction	£30–600 t^{-1}
Chemical dehalogenation	£150–420 t^{-1}

stage to an established technology. During time the understanding of performance is likely to improve, and as techniques become established they acquire a better track record and become more available. Established process-based methods are essentially restricted to incineration, solidification/stabilization, biological treatments, soil washing, thermal desorption and some *in situ* processes.

Availability

The availability of particular remediation processes will depend to some extent on the development status of the process and also on the market demand for it. When considering an appropriate technology it is normal to examine case studies

[10] Royal Commission on Environmental Pollution, *Sustainable Use of Soil*, Cm 3165, HMSO, London, 1996.

[11] P. Crowcroft, P. Bardos and P. Wood, in *Polluted + Marginal Land 92; Proceedings of Second International Conference on Polluted and Marginal Land*, Engineering Technics Press, Edinburgh, 1992, pp. 267–286.

[12] M. Harris, paper presented to Cycle de Conferences sur l'Environnement, Societe Royale Belge des Ingenieurs et des Industrials, Brussels, March 1991.

where the process has been used successfully before. If a technology has only had restricted prior use then this, in turn, may limit its availability, as vendors experience difficulties in marketing untried processes. Consequently, although some processes are available and have been used overseas, their availability and application in the UK may be considerably less.

Operational Requirements

Operational requirements to be considered during process selection include all measures and activities necessary to undertake the remediation. Operational requirements will vary considerably and may result in a technically viable and cost effective remediation approach being rejected. Operational aspects to be considered include:

- health and safety requirements
- legal/regulatory issues
- access and transport issues
- infrastructure requirements
- environmental protection
- time constraints
- quality assurance requirements

Information Requirements

An appropriate remediation strategy can only be determined as a result of appropriate site information. The information and data required can be extensive and, although much may be available from site investigation/risk assessment activities, it is possible that additional information will be required. It should be recognized that the information obtained to determine the presence of a contamination problem and the need for information is generally not sufficient to determine the appropriate remediation approach.

Monitoring Needs and Potential Environmental Impact

Monitoring may be necessary during remediation in order to provide:

- quality assurance
- process control and optimization
- environmental protection
- compliance with health and safety requirements

Additionally, further monitoring may be necessary following completion of the remediation and so may need to be considered as part of the post-treatment management strategy.

The monitoring regime required will depend on the type of remediation process being applied, site-specific factors and the types of hazards and associated risks that might be present. For example, stringent monitoring of the quality of imported materials may be necessary if a cover system is being installed. These materials need to be contaminant free, clays need to conform to the specified permeability, and individual components of a cover system need to be constructed to specification thickness. Process-based remediation methods will require monitoring to verify that treatment targets have been achieved. With *in situ* processes, considerable monitoring may be necessary to ensure containment of any reagents or other additives associated with the process and to determine that the remediation has been successful throughout the entire contaminated area. Additionally, if *in situ* bioremediation processes are being used, it will be necessary to monitor the occurrence and mobility of any hazardous intermediate or by-products. Other environmental requirements may extend to monitoring of contaminated dusts or particulates, noxious gases and liquids or any waste streams.

Health and Safety Needs

All remediation methods require some degree of health and safety provision during their execution. However, requirements will depend on the hazards presented by the contaminants present on any site, and the type of remediation being employed. The UK Health and Safety Executive has produced guidance for the protection of workers and the public during the development of contaminated sites[13] and also a discussion of occupational hygiene aspects associated with different treatment types.[14]

Post-treatment Management Needs

Post-treatment management regimes may be necessary either to verify the success of the remediation process or to monitor the long-term integrity of the remediation. Verification has been considered above. Longer-term monitoring may be required for a range of circumstances, including monitoring the integrity of cover or containment systems to ensure that no contaminated leachate escapes or that any gases are being adequately dealt with, and monitoring of stabilized/solidified products for long-term deterioration. Harris[8] outlines those aspects that need to be considered during the development of a verification programme.

13   Research Trends

A significant amount of research into remediation processes for contaminated ground is currently being undertaken. This includes basic research in the form of experimental or theoretical work aimed at obtaining new knowledge for the understanding of processes and environmental observations, and applied research directed towards a specific practical objective.

Structured research programmes relevant to bioremediation processes include

---

[13] Health & Safety Executive, *Protection of Workers and the General Public During the Development of Contaminated Land*, HMSO, London, 1991.

[14] Health & Safety Executive, *Remediation of Contaminated Land, Occupational Hygiene Aspects on the Safe Selection and Use of New Soil Clean-up Techniques*, Specialist Inspector Reports No 51, Health & Safety Executive, Sudbury, 1996.

the Environmental Biotechnology Programme funded by the UK Biotechnology and Biological Sciences Research Council, the LINK Biological Treatment of Soil and Water Programme, and various UK research clubs. Research is often directed towards the treatment of specific contaminants, in specific media, and by specific biological agents. Contaminants being investigated include pesticides, spent oxides/cyanides, petroleum hydrocarbons, PAHs and PCBs.

Research into the physical treatment of contaminated soil includes investigations into soil separation and washing within the UK Department of the Environment's Contaminated Land Research Programme, and full-scale remediation of contaminated ground and groundwater using soil vapour extraction, pump-and-treat, air stripping and carbon adsorption by AEA Technology. Other research includes investigations into hydraulic control and pump-and-treat, and electroremediation.

Research into stabilization/solidification and chemical and thermal processes is more restricted. Investigations into the use of stabilization/solidification include examination of various types of products incorporating modified organophilic clays, lime, fly ash and zeolites. Investigations into chemical processes are concentrating mainly on chemical leaching and extraction, and oxidation. Research into thermal processes is concentrating on organic contamination.

Further information on future research trends has been summarized by Martin and Bardos.[15]

14 Conclusions

The primary aims of remediation of a contaminated site are either the reduction of actual or potential environmental threat, or the reduction of potential risks that are unacceptable to levels that are acceptable. Remediation is achieved by undertaking one or more of the following: removal or destruction of the contaminants, modification of the contaminants to a less toxic, mobile or reactive form, or isolation of the contaminant from the target by interrupting the pathway of exposure.

Remediation methods can be grouped into two broad approaches: engineering methods and process-based techniques. Engineering methods include excavation and disposal to landfill, and the use of containment methods. Process-based techniques include physical, biological, chemical, stabilization/solidification and thermal processes. These can be used as stand-alone treatments or can be combined to form treatment trains. Although many process-based techniques are at various stages of development, the number of technologies that have been proven or that are widely available is considerably less. Established processes that are available in the UK are essentially restricted to incineration, thermal desorption, solidification/stabilization, biological treatments, soil washing and some *in situ* processes.

Identifying the most appropriate method for the remediation of a given site is a difficult process and requires consideration of a number of factors. These include

[15] I. Martin and P. Bardos, *A Review of Full Scale Treatment Technologies for the Remediation of Contaminated Soil*, BPP Publications, Richmond, 1996.

process applicability, effectiveness and costs, process development status and availability and operational requirements. Additional factors to be considered are process limitations, monitoring needs, potential environmental impact, health and safety needs and post-treatment management requirements. The amount of information that is needed for an effective appraisal of available options is considerable and may, in many situations, not be available. It is pertinent to note that the details of a site investigation needed to determine that a site is contaminated and requires remediation is generally not sufficient to identify what remediation approach would be the most effective.

During any remediation process, adequate quality control measures are needed to ensure that the methodology conforms to specification or that treatment targets have been achieved. In many cases this is likely to require environmental monitoring while remediation is in progress. Additionally, upon completion of remediation, additional monitoring and management activities may be necessary, especially if any contamination remains in any form at the site.

Land Reclamation after Coal-mining Operations

DAVID L. RIMMER AND ALAN YOUNGER

1 Introduction

Coal fuelled the 19th century Industrial Revolution in Europe and North America and was also the major source of energy for industry and domestic use for much of the 20th century. The mining of that coal was mainly from underground operations which produced large amounts of waste. The disposal of those wastes was generally not controlled, and was done by creating a waste heap adjacent to each mine shaft. These heaps were often conical in shape from tipping out of overhead conveyors. As these tips grew they became prominent features in the landscape; in the UK they acquired a range of popular names, including pit heaps, slag heaps and bings. The material in them is technically referred to as colliery spoil in the UK, and hence the accepted technical description of the tips is colliery spoil heaps.

Starting in the 1950s the development of large earth-moving equipment changed the economics of coal mining, such that deep underground mines became progressively less profitable and were gradually closed, and coal was increasingly mined from the surface by what became known in the UK as opencast mining, and in the USA, more graphically, as strip mining. These changes created two problems: increasing numbers of abandoned deep mines with their associated spoil heaps, and large areas of land being disturbed by opencast mining. For the latter, legislation required that the land be restored to its former use (or an alternative if appropriate) after mining ceased. The technical problems in achieving successful restoration of such land required a considerable research effort, and that is reviewed in Section 3. The problem of the abandoned deep mines and the reclamation of the sites affected by them is covered in Section 2. It is somewhat ironic that the technology of large earth-moving machinery, which caused the demise of underground coal mining and the problem of abandoned deep-mine sites, should have been crucial in providing the means for their reclamation.

In the remaining parts of this Introduction the broad problems faced in the reclamation of land following coal-mining operations are described, and the key role of the subsequent land use in determining the reclamation procedures is discussed.

73

D. L. Rimmer and A. Younger

Coal Mining, Derelict Land and Land Disturbance

The creation of a deep mine inevitably involved the production of much rock waste from the sinking of shafts and the digging of underground passages to reach the coal seams. This rock waste was brought to the surface and dumped near the pit-head. With time, the heaps of this material came to occupy areas of land proportional to the volume of the underground workings. The land on which the heaps stood was permanently affected, because the mining operators were under no obligation to remove the material or treat it in any way. When the mines were closed these areas became officially designated as 'derelict land', defined by the UK government as 'land so damaged by industrial or other development that it is incapable of beneficial use without treatment'.[1] In other words, the land needed to be reclaimed.

The costs of reclamation were usually very great and, as there was no obligation on the mine owners to pay these costs, the land generally passed into public ownership and the cost of reclamation was borne by central government, *i.e.* by the taxpayer. The rationale for carrying out the reclamation was to reduce the impact on the landscape of these unsightly features, which symbolized industrial decline. The heaps were often slow to revegetate naturally;[2] the bare surfaces would dry out in the summer and could lead to a dust problem. In addition, because of the varying amounts of residual coal that they contained, many heaps caught fire and this was obviously hazardous.[3] So the reclamation was in effect an environmental clean-up operation, with the ultimate end-use of the land less important than the fact that the spoil heaps had been removed or made less prominent features in the landscape.

It is interesting to note that the broad definition of derelict land given above could equally well be applied to much 'contaminated land'; this term has come to be used specifically for land affected by chemical substances. It has been estimated that 65% of the land classified as derelict in the 1988 survey was potentially contaminated.[4] The terminology used to describe the treatment of such land to bring it back into use is different in the two cases. Derelict land in general undergoes 'reclamation', whereas 'remediation' is carried out on contaminated land.

In contrast to the creation of derelict land by the deep mining of coal, surface or opencast mining only leads to temporary land disturbance. There is an obligation on operators to restore the land when mining ceases. For this reason the process of reclaiming the land is usually referred to specifically as 'restoration' rather than as reclamation. The excavation of soil and rock overburden in order to reach the coal seams sometimes leads to the creation of large temporary storage heaps of soil and rock materials. These are subsequently replaced, but the soil is substantially disturbed in the process and the restored land requires prolonged and careful treatment to bring it back to its former state.

[1] Dept. of the Environment, *Survey of Derelict Land in England* 1988, HMSO, London, 1991.

[2] J. A. Richardson, B. K. Shenton and R. J. Dicker, in *Landscape Reclamation*, ed. B. Hackett, IPC Science and Technology Press, Guildford, 1971, vol. 1, p. 84.

[3] G. P. Doubleday, *Outlook Agric.*, 1974, **8**, 156.

[4] Royal Commission on Environmental Pollution, 16th Report, Cm 1966, HMSO, London, 1992, para 7.57.

Reclamation and After-use Choice

For the reclamation of colliery spoil heaps the subsequent land uses can include agriculture, forestry, nature reserves, sports fields, golf courses, public open space, housing or industrial use. Most of these require an element of revegetation; in some cases that may be relatively restricted, such as low maintenance grass cover on sports fields, golf courses and public open space. In other cases such as agriculture, forestry and nature reserves, there is a need to provide a much better medium for plant growth in order to maximize the productivity for agriculture and forestry, or to tailor the plant growth characteristics for specific nature conservation objectives. Because agriculture and forestry have been the most common land uses for reclaimed colliery spoil and because of their demanding plant growth requirements, they have received the most research attention. For the purposes of this article we will concentrate on the agricultural after-use, with which we are most familiar. However, many of the same principles apply in forestry. Similarly for restored opencast land, the commonest after-use is agriculture and we will concentrate on that in Section 3.

2 Reclamation of Deep-mine Spoils

Deep-mine Colliery Spoils

Deep-mine spoils are the rock wastes brought to the surface during the mining operation. They are composed mainly of the rock types found adjacent to the coal seams; this means shale and, to a lesser extent, sandstone, together with some coal not separated during processing. The particle size distribution of the material on the surface of the resulting spoil heaps is dominated by stones (*i.e.* material with diameter greater than 2 mm). The data in Table 1 are for three sites sampled immediately prior to revegetation.[5] The material was disturbed during site preparation (see below) and this, together with weathering, may have led to some comminution; even so, the stone content ranges from 50% to 75%. Clearly this is not an ideal plant growth medium, added to which the remainder of the material (the soil-sized, <2 mm, fraction) is dominated by clay and silt, which again is not ideal. This fine-grained material is produced when shale is weathered.

In some heaps, combustion of residual coal leads to major changes.[3] Temperatures up to 1000 °C have been recorded in burning heaps, and this leads to the oxidation of all organic material and also any reduced iron compounds,

[5] D. L. Rimmer and P. Colbourn, *Problems in the Management of Soils Forming on Colliery Spoils*, Report for Dept. of the Environment under research contract DGR 482/7, 1978.

Table 1 Particle size distribution (%) of the 0–10 cm layer at three colliery spoil reclamation sites[5]

	Tudhoe	Ashington	Abertysswg
Stones (>2 mm)	51.3	61.0	75.8
Sand (0.02–2 mm)	11.5	11.9	8.2
Silt (0.002–0.02 mm)	17.0	12.4	7.2
Clay (<0.002 mm)	20.2	14.6	8.8

such as iron(II) sulfide (FeS_2, iron pyrites). The most obvious result is a change in the overall colour of the spoil from a dull grey to lighter and brighter colours from cream to red. The temperatures are also sufficient to create large fused masses of material, up to 10 m in diameter. Such 'burnt spoil' is chemically less reactive and physically stabilized, and is used in construction, for example as a base in road-building, in preference to the unburnt spoil.

Apart from those brought about by combustion, changes to the material while stored in the waste heaps will generally be quite small; those due to physical and chemical weathering processes will be limited to the top few metres. Physical weathering will be most effective on the shales, which readily break down and release clay- and silt-sized material. Chemical weathering is particularly pronounced when iron(II) sulfide is present, as this oxidizes to iron(III) and sulfate and generates considerable acidity. FeS_2 is not present in all colliery spoils, and when it is present in a heap its distribution will be patchy, reflecting the changing geological strata excavated underground and then brought to the surface. This highlights the fact that there is considerable heterogeneity within colliery spoil heaps, as well as differences between them.[6]

The pH of the material is generally in the range 5–6.5, except in the presence of oxidizing iron pyrites, when the value can fall to 3 or less. The normal range is satisfactory for grass growth, but there are serious deficiencies of major plant nutrient elements, particularly nitrogen and phosphorus.

In addition to the excavated materials, the waste heaps often contain other non-geological materials which have been dumped there. This rubbish includes bits of abandoned machinery, conveyor belts and steel hawsers. Where this is near the surface it is removed during site preparation.

Site Preparation

Colliery spoil heaps form prominent and obtrusive features in the landscape. One of the objectives in their reclamation is to make them blend in with their surroundings more effectively. To achieve this they need reshaping to produce less steep slopes and smoother contours, a process known as regrading. During regrading it is possible to bury well below the surface any material with a large iron sulfide content, because this has the potential to oxidize in the presence of air and water and generate acidity in the process. Unless large amounts of material are removed from the site, which is rarely the case, then the regrading involves spreading the waste over adjacent land. The soil, both topsoil and subsoil, from any such adjacent areas is stripped and stored before regrading begins. This stored soil is a valuable material for use in the subsequent revegetation.

The heaps are regraded by very heavy earth-moving machinery, and this leaves the material at the surface in a very compacted state. To alleviate this compaction and at the same time bring to the surface any large stones or rubbish, the next operation is 'ripping'. This is done with rigid metal tines 60 cm in length, set 60 cm apart, which are pulled through the material. Any stones and rubbish are then cleared from the surface of the site.

[6] A. H. Fitter, J. F. Handley, A. D. Bradshaw and R. P. Gemmell, *Landscape Design*, 1974, **106**, 29.

Revegetation without the Use of Soil

The establishment of a grass sward is usually the primary objective of revegetation for agricultural after-use. On most sites this can be achieved by following normal agricultural practices: cultivation with shallow tines to create a seedbed, application of nitrogen and phosphorus fertilizers, harrowing to incorporate the fertilizers, spreading of a seeds mixture, followed by rolling. Typical specifications for these operations are given in Table 2.[7] The seeds mixture usually contains perennial ryegrass, one or more varieties of clover and other grasses such as fescue, timothy and cocksfoot. A typical mixture, applied at a rate of 56 kg ha^{-1}, is perennial ryegrass (*Lolium perenne* varieties: S23, 33 kg and S101, 11.5 kg), timothy (*Phleum pratense* variety S48, 6.5 kg), white clover (*Trifolium repens* variety S100, 2.75 kg) and wild white clover (*Trifolium repens* variety S184, 2.75 kg).[5] Establishment is usually successful, except on areas affected by acid production where pH values may fall to 4 or less. The presence of bare patches on a revegetated site is usually a good indication of acid conditions.

In the first season, productivity of such grassland can be as good as on undisturbed agricultural land, with dry matter yields up to 10 t ha^{-1}. At this stage the soil conditions, both physical and chemical, are optimized. Thereafter there is a deterioration, which results in progressively poorer growth of grass in subsequent seasons. The rate of decline of yield has been measured on experimental plots at an average of 1.5 t ha^{-1} a^{-1} over the first five years.[8] The cause of this decline is largely the deteriorating soil physical conditions, due to settling of the spoil. This leads to compact material in all but the top 5 cm.[9] The symptoms of this compaction are poor water-holding, which leads to drought in the summer, and poor drainage, which causes waterlogging in the winter. The effect of these adverse conditions is not only to reduce grass yield, but also to cause changes in the sward composition, with species better adapted to the conditions, *e.g.* cocksfoot, increasing at the expense of other less well-adapted species, *e.g.* ryegrass.

A survey of the physical conditions at 12 sites in north-east England showed that bulk density in the rooting zone from 5 to 30 cm depth was approximately 1.8 g cm^{-3} and total porosity less than 20% (v/v).[9] Greenhouse experiments demonstrated that there was a threshold density in the region of 1.4–1.5 g cm^{-3}, above which growth of ryegrass was reduced.[10] Increasing compaction had more effect on roots than on shoots, because the root:shoot ratio decreased with increasing compaction.

To overcome these problems and maintain a productive sward for grazing, it is necessary to cultivate and reseed areas which are suffering from compaction. The cultivation necessary can be very shallow. In field experiments it was found that discing or spring-tine cultivation to 5 cm was as successful as deeper cultivations in terms of the grass yield in the seasons immediately following.[5]

[7] G. P. Doubleday and M. A. Jones, in *Landscape Reclamation Practice*, ed. B. Hackett, IPC Science and Technology Press, Guildford, 1977, p. 85.
[8] D. L. Rimmer and A. Gildon, *J. Soil Sci.*, 1986, **37**, 319.
[9] D. L. Rimmer, *J. Soil Sci.*, 1982, **33**, 567.
[10] D. L. Rimmer, *J. Sports Turf Res. Inst.*, 1979, **55**, 153.

Table 2 Operations typically carried out in establishing a grass sward on reclaimed colliery spoil[7]

Operation	Details
Liming	If needed, apply ground limestone at a rate not exceeding 10 t ha^{-1}
P fertilizer	Apply basic slag (or equivalent) at a rate not exceeding 2.5 t ha^{-1}
Primary cultivation	Chisel plough to a depth of not less than 150 mm; repeat at 30° to the first pass
Site clearance	Remove all stones, bricks, wire, wood or other hard objects exceeding 50 mm in any dimension. Cables and wire shall, if possible, be pulled out; if this is not possible, they shall be cut off at least 250 mm below ground level
Grading	Scrub to leave the ground surface with true and even grades. This operation shall be undertaken only during dry ground conditions
Seedbed preparation	Chain harrow and Cambridge roll until a fine firm tilth is obtained
N/K fertilizer	Spread fertilizer at a rate not exceeding 0.5 t ha^{-1}. Light harrow
Seeding	Supply and broadcast the seed mixture as specified for the area. The seed shall be applied in two passes, the second at right angles to the first, half of the seed being spread in each pass. Cambridge roll
Final site clearance	Remove any stones, bricks, wire, *etc.*, of greater than 50 mm dimension

Many of the same problems arise when the post-reclamation land use is forestry.[11] In establishing trees the main cause of poor success is physiological drought, which is brought about by the small water-holding capacity of the colliery spoil and the competition with ground cover vegetation for water. Similarly, growth of established trees is controlled principally by water availability. In order to overcome these problems it is recommended that, in both planting and aftercare, provisions are made to keep the area around the trees free from ground cover.[11]

A related factor in reclamation to forestry is compaction, which can be a major problem for both establishment and growth. Cultivation to reduce the soil bulk density can be undertaken before planting, but the effects can be short-lived.[12] Selection of suitable species of trees that are best adapted to the conditions of summer drought, winter waterlogging, stoniness and compaction can help. The species recommended are alders, birch, hawthorn, ash and pine (Lodgepole, Scots and Corsican).[12]

[11] R. N. Humphries and P. R. Benyon, *Investigation of Tree Planting on Restored Colliery Tips and Lagoons*, British Coal Corporation, Eastwood, Nottinghamshire, 1994.

[12] Dept. of the Environment, *The Reclamation of Mineral Workings*, Minerals Planning Guidance Note 7 (revised), HMSO, London, 1996.

Revegetation Using Soil

Because of the problems encountered in revegetation directly on the colliery spoil material, it is now generally recognized that the use of soil is desirable, if at all possible. At many sites it is possible to obtain soil from adjacent, unaffected land, which is incorporated into the reclamation scheme (see *Site Preparation* above). In other cases it must be imported on to the site. In either case the quantities available are likely to be quite small and it is therefore important to use it as effectively as possible. The questions that arise therefore are: what is the minimum amount of soil which will be effective in different situations; and is that soil more effective placed as a cover layer on the spoil material or incorporated as an amendment in the spoil?

To answer these questions a number of experimental trials were carried out at two sites in the north of England.[13] In one set of trials, soil layers of increasing thickness from 5 to 25 cm were applied as a cover, and a grass-clover sward established. In general there was a benefit in terms of grass yield from increasing the depth of soil cover up to 15 cm. However, the effects were more noticeable at the site which had an acidifying spoil material, owing the presence of iron pyrites. The effect of the soil cover in that case was to reduce the extent of acidification in the surface material (Table 3). Thus the more severe the conditions in the underlying spoil, the more benefit there was from the soil cover. On such sites there would be no benefit from incorporating the soil, because its function as a barrier would be destroyed. On less acidic or non-acidic spoil, soil has less benefit in the short term, but from earlier evidence it maintains productivity in the longer term, because it improves physical and biological conditions.[8,14] The experiments on soil incorporation were only carried out at the non-acidic sites. The results showed that overall there was a small benefit from incorporation, but that it was probably too small to justify the extra effort and cost.[13]

In field trials in which various methods of reclamation were tested, the long-term beneficial effect of soil as an amendment was established.[8] The improved conditions for plant growth were found to be chemical, physical and biological. In order to quantify the biological benefits at a larger number of sites, a survey was carried out at a number of soil-covered reclamation sites in the north of England.[14] The soil respiration (CO_2 evolved) was measured as a gross indicator of biological activity, along with depth of soil cover and a range of chemical and physical properties likely to affect biological activity. The survey showed that depth of soil cover was not an important factor affecting biological activity, except in cases where the soil was covering highly acidic spoils. In such cases a greater depth of soil led to greater surface soil pH and hence to greater soil respiration, in agreement with the field trails described above.[8] Soil organic matter content was a very important factor influencing biological activity, its effect being a combination of a chemical or nutritional one, in terms of N and C supply to soil organisms, and a physical one, by influencing water-holding capacity and bulk density. Stepwise multiple regression analysis showed that water-holding capacity was the most important single

[13] A. Gildon and D. L. Rimmer, *Soil Use Manage.*, 1993, **9**, 148.
[14] A. Gildon and D. L. Rimmer, *Biol. Fertil. Soils*, 1993, **16**, 41.

Table 3 Effect of thickness of soil cover on pH of the surface material and buried spoil[15]

	Soil thickness (cm)						Least significant difference (P = 0.05)
	0	5	10	15	20	25	
Surface material pH							
Start of experiment	5.9	6.5	6.5	6.5	6.5	6.5	—
End of experiment	4.1	5.7	5.8	6.0	6.2	6.3	0.2
Buried spoil pH*							
End of experiment	—	3.9	4.8	4.4	4.4	4.6	1.2

*Material from just below the soil:spoil boundary.

factor, suggesting that, on the sites surveyed, physical factors were more important than chemical ones.

3 Reclamation of Opencast Coal-mining Sites

The technique of opencast mining involves temporary removal of the soil and the associated rock strata (overburden) to expose coal seams which are relatively close to the surface. At many UK mines, individual coal seams may be relatively thin, but several seams will be mined during the working of an individual site. As time has progressed, and the machinery technology has advanced, the depth to which profitable opencast mining can be carried out, and the ratio of earth moved to coal extracted, have both tended to increase substantially. Whether small sites were being proposed by private mining contractors, or large sites were being planned by the previously nationalized British Coal, the planning procedures relating to opencast mining in the UK have always included stipulations about the eventual end use of the site, and the condition of the land after the mining and restoration processes were completed. Officers of the Ministry of Agriculture, Fisheries and Food (MAFF) had to be satisfied that the obligations of the mining contractor had been fulfilled before land could be released for farming or other uses.

In the early days of opencast mining, the standards set were relatively modest and could easily be met, leading to some poor quality restorations after which the productivity of the land was substantially reduced, and its use was restricted to relatively extensive grassland or forestry. As time has progressed, there has been a considerable increase in the expectations from restored land, and there has been a gradual improvement in restoration standards with a wider range of vegetation management options now considered possible.[15]

The Problems

The processes of removing, storing and subsequently replacing the soil during the mining activity each lead to potential problems in relation to subsequent restoration. In this respect, a major distinction should be drawn between those

[15] J. A. King, *Soil Use Manage.*, 1988, **4**, 23.

sites where, for operational reasons, soil has to be stored for a period of years while the mining progresses and those, usually larger, sites where a progressive system of restoration can be practised. In this latter instance, the topsoil (30 cm) and subsoil (variable depths depending on site) are removed in sequence from the part of the site which is to be mined next, and taken directly to the part of the site where mining has been completed and finished contours have been established using overburden. There it is re-laid in sequence without undergoing a period of storage.

The problems with progressive restoration are principally associated with the damage to soil structure caused by lifting and replacing the soil profile using heavy earthmoving equipment. The type of machine most frequently used for this purpose, the box scraper, tends to cause smearing of soil at the surfaces where the blade cuts through the soil, and compaction when the soil is forced into the box by the forward motion of the vehicle and similarly when it is released, and when each successive layer is travelled over by the machine as it lays the next layer.

Measurements have consistently shown that the dry bulk density of the soil is increased by these operations (Table 4), pore space is reduced and the movement of excess water through the restored profile is slower than that in an equivalent unworked soil.[15] However, apart from small effects attributable to inadvertent mixing of top and subsoil, the chemical properties of the soils, and therefore the availability of most major plant nutrients, remains unaffected (Table 5).[16] Whilst the microbial populations in the soil may be little affected, there is evidence that the populations of larger organisms, *e.g.* earthworms, are adversely affected. Martin suggested that mortality rates of more than 90% were observed each time the soil was moved, because organisms may be buried, cut, crushed or exposed to predation by birds.[17] It has also been observed that there was a disproportionate decrease in the numbers of the larger, deep burrowing species, *Aporrectodea longa* and *Lumbricus terrestris*;[18] the beneficial effects of these organisms in terms of burrowing and recycling of organic matter within the soil profile are therefore greatly reduced.

When the mining operations also involve the storing of the soil for a period of time, additional problems arise because of the anaerobic conditions which occur below the surface in the soil storage mounds. Populations of obligate aerobic fungi and actinomycetes are substantially reduced in the anaerobic part of the mound (below 1 m). Numbers of bacteria decline less rapidly with depth, indicating that some are facultative anaerobes capable of adjusting to anoxic conditions at depth. Large amounts of ammonia begin to accumulate at depth in the stockpiles as a result of the anaerobic decay of soil organic matter.[19,20] When previously stored soils are respread, nitrate production from this ammonia proceeds rapidly in the aerobic conditions, placing at risk very large amounts of labile nitrogen in the short term[20,21] and leading to much lower levels of available

[16] E. J. Evans, M. H. Leitch, R. I. Fairley and J. A. King, *Reclam. Reveg. Res.*, 1986, **4**, 223.

[17] A. D. Martin, R. N. Humphries and W. J. Whittington, in *A Seminar on Land Restoration Investigation and Techniques, University of Newcastle upon Tyne*, British Coal Opencast Executive, Mansfield, UK, 1988, p. 72.

[18] W. C. O'Flanagan, G. J. Walker, W. M. Waller and D. Murdoch, *NAAS Q. Rev.*, 1963, **62**, 85.

[19] J. A. Harris, P. Birch and K. C. Short, *Soil Use Manage.*, 1989, **5**, 161.

[20] D. B. Johnson, J. C. Williamson and A. J. Bailey, *J. Soil Sci.*, 1991, **42**, 1.

[21] R. Davies, R. Hodgkinson, A. Younger and R. Chapman, *J. Environ. Qual.*, 1995, **24**, 1215.

Table 4 A comparison of the physical properties of a restored and an unmined soil in the north-east of England (Dunkeswick soil series)[15]

Site	Topsoil		Subsoil	
	Unmined	Restored	Unmined	Restored
Stone content (%)	1.8 (0.30)	1.7 (0.24)	1.9 (0.40)	4.6 (1.63)
Dry bulk density ($t m^{-3}$)				
Whole soil	1.29 (0.38)	1.38 (0.032)	1.46 (0.053)	1.64 (0.037)
Fine earth	1.27 (0.041)	1.36 (0.033)	1.45 (0.051)	1.59 (0.024)
Total porosity ($m^3 m^{-3}$)				
Whole soil	0.48 (0.014)	0.46 (0.010)	0.45 (0.018)	0.37 (0.017)
Fine earth	0.48 (0.015)	0.45 (0.015)	0.45 (0.020)	0.37 (0.012)

Figures in parentheses are standard errors.

Table 5 Comparison of some soil fertility parameters from an unmined site and a restored site in the north-east of England (Dunkeswick soil series)[16]

	Site	
	Unmined	Restored
pH	6.41 (0.05)	6.87 (0.08)
CEC (meq $100 g^{-1}$)	15.01 (0.44)	12.53 (0.52)
Ca ($mg kg^{-1}$)	2345 (66.4)	2794 (100.6)
Mg ($mg kg^{-1}$)	244 (19.7)	225 (14.1)
K ($mg kg^{-1}$)	84 (3.9)	89 (6.4)
Na ($mg kg^{-1}$)	28 (1.5)	39 (2.2)
Mn ($mg kg^{-1}$)	11 (0.9)	9 (0.9)
P ($mg kg^{-1}$)	17 (1.2)	7 (0.7)
Organic carbon ($g kg^{-1}$)	35.7 (0.98)	35.5 (1.06)

Figures in parentheses are standard errors. CEC = Cation exchange capacity.

nitrogen in subsequent years. These chemical and microbiological changes therefore add to the physical problems associated with the movement of soil, which in the case of stored soils is handled by the damaging machinery on at least twice as many occasions as progressively restored soil.

Overcoming the Problems

Because of the problems of the horizontal layering of replaced soil profiles and the compaction of each layer, crop rooting is often restricted and percolation of water into the soil is limited. Typically, rain falling on recently restored soils will wet the surface and then, failing to percolate, will either remain at the surface as waterlogging or run off the surface, leading to rapid water loss and failure to recharge the soil moisture below the surface, as well as increasing the possibility of erosion on some of the more sloping land. All these effects on the soil will also manifest themselves as poor crop establishment and growth, whether of agricultural grassland or other forms of vegetation. The extent to which the above problems will occur will depend on the initial soil type, on the local climate, on the methods used to move the soil and on the condition of the soil at the time of its movement. Many coal resources, *e.g.* those in the north-east of England, are associated with relatively heavy land of high clay content which in normal circumstances is slowly draining. In the East Midlands there are sites where the soil contains a greater proportion of sand and is inherently better drained. In both of these areas the annual rainfall is modest, so the need for surplus water control is substantially less than in the South Wales coalfield, where the combination of high annual rainfall, silty clay soils of inherently poor aggregate stability and often quite steeply sloping land is most difficult. In all instances, problems of soil structure are exaggerated if the soil has been moved when conditions were unsuitable. In particular, soils should be moved when they are in a friable state and not in wet conditions, when they are at, or above, the plastic limit, because of the excessive smearing and compaction that can follow. In England and Wales, MAFF officials have the authority to prevent contractors from carrying out soil movement operations under such unsuitable conditions.

Installation of a Drainage System

In order to begin the process of redevelopment of soil structure, the installation of an effective underdrainage scheme is considered vital. This normally comprises two stages. First, the installation of a system of underground pipes at intervals across the area. The void above the pipe is back-filled with gravel to the surface. Second, the use of secondary drainage treatments such as subsoiling or mole draining to improve the flow of water into the piped drains. Investigations of drainage designs[22-24] have generally shown that the spacing between drains is not critical, but that designs based on agricultural criteria are usually inadequate to cope with peak flow rates which occur after storms. Basically, because of the limited percolation of water into the soil on restored sites, water simply flows along the surface or through deep fissures, to be collected in the drains and removed from the site. The rate and volume of flow are therefore both greater and more rapid than on unworked soils, where the soil acts as a reservoir for water, and pipes may be unable to cope with peak rates of flow in the short term, so the system becomes surcharged. As an example of this, Table 6 shows the frequency with which drain flow exceeded design flow in a drainage experiment in the north-east of England over a four-year period.

Secondary drainage treatments are therefore recommended in order to increase the infiltration of water into the soil and to conduct surplus water more effectively from the surface to the piped drain. After the installation of the piped system, it is

[22] R. A. Hodgkinson, N. C. Bragg and R. Arrowsmith, in *A Seminar on Land Restoration Investigation and Techniques, University of Newcastle upon Tyne*, British Coal Opencast Executive, Mansfield, UK, 1988, p. 3.

[23] R. A. Hodgkinson, Research Paper 6, ADAS Field Drainage Experimental Unit, Trumpington, Cambridge, 1988.

[24] J. Scullion and A. R. A. Mohammed, *J. Agric. Sci., Camb.*, 1986, **107**, 521.

Table 6 Number of drainage events per annum exceeding the 1 in 2 year design* flow at Butterwell restored site, Northumberland[23]

	Drains at 20 m spacing with mole drainage and subsoiling		Drains at 40 m spacing with mole drainage alone	
Plot	1	2	1	2
1982/83	33	32	29	23
1983/84	8	8	15	14
1984/85	23	28	24	40
1985/86	21	31	27	29

*The drainage scheme was designed to cope with the removal of surplus water from rainfall events of an intensity which is likely to occur once every two years.

now normal either to break up the surface of the soil using straight or winged tine subsoilers and/or, if the soil is of a suitable clay-rich texture, to introduce mole drains at an angle to the piped system. Results from experiments measuring the effects of such treatments have been variable. For example, on a reasonably well-structured progressively restored site in the north-east of England, several secondary drainage treatments failed to produce any effect on the soil or the associated crops relative to an uncultivated control treatment (Table 7),[25] whereas on a site in the same region where an initially poorer soil had been replaced after storage, penetration resistance of the soil and crop root growth were both improved, leading to higher crop yield and nutrient uptake. In one instance, the use of a subsoiler on an unstable soil increased the infiltration of water into the surface of the soil, reducing its bearing strength and leaving the area boggy and unmanageable.[22] Most workers have commented that any positive observed effects of subsoiling and moling are often relatively short lived, because soils may slump with time, and fissures or channels previously created tend to fill up with fine particles. Thus, on some suitable soil types, and especially where the agricultural management allows for regular cultivation, repeated subsoiling can be advised in order to continue the benefit. Even on silty soils in high rainfall areas, where the only reasonable crop would be long-term grassland, regular subsoiling to maintain contact between the surface and the piped drainage system has been recommended. There it has been recognized that subsoiling through an established pasture reduced surface wetness and increased earthworm abundance, and although this did not affect production or crop rooting immediately, it would contribute to land rehabilitation in the broader sense.[26]

The timing of the installation of the drainage scheme has also been investigated. Originally it was considered prudent to wait for two or three years after soil replacement before installing the drainage scheme, principally to allow for any settlement to occur. However, experiments have shown that crop root growth, nutrient uptake and yield are reduced in the period during which the land remains undrained, and little or no development of soil structure occurs during this period. Table 8 shows the yield of barley crops grown at the Butterwell site in

[25] A. Younger, *Soil Use Manage.*, 1989, **5**, 150.

[26] J. Scullion, *Restoring Farmland after Coal*, British Coal Opencast Executive, Mansfield, UK, 1994.

Table 7 Effects of subsoil cultivation technique on yields of grass herbage and wheat at a restored site (Butterwell)[25]

		Straight tine	Winged tine	Mole	Control
Herbage yields	1986	10.5	10.4	10.3	10.6
(t DM ha^{-1})*	1987	8.8	8.1	8.1	8.4
Wheat yields	1986	6.7	6.6	7.0	6.8
(t ha^{-1})	1987	7.4	7.0	7.1	7.1

*DM = Dry matter.

Table 8 The effect of date of drain installation on the grain yield (t ha^{-1}) of barley crops grown on two parts of a restored site

Harvest year	Site 1 (Drains installed in 1981)	Site 2 (Drains installed in 1986)
1982	7.82	5.64
1983	4.82	3.00
1984	5.01	3.43
1985	4.75	1.81
1986	—	—
1987	7.4	7.1
1988	7.8	8.2

the north-east of England on drained and undrained restored land, and the effect of introducing a drainage scheme on the previously undrained area. The positive effect of drainage is clearly apparent, and it occurred immediately after the installation of the scheme. Thus, there has been a trend in more recent restorations towards installing a drainage scheme as early as other site considerations will allow.

Choice of Crop and its Management

Having installed a drainage scheme, the question arises as to how best to manage the site agriculturally in order to complete the most effective restoration. In the UK this is particularly relevant during the five years after soil replacement, when the management is under the control of the Agricultural Development and Advisory Service (ADAS). During that period in particular, cropping need not be geared towards maximum profit, but can be designed to maximize soil structure development. Traditionally, grass has been seen as the favoured crop during the early stages of restoration because of the well documented effects of the fibrous root system on soil structure. However, it might be argued that green manure crops, regularly ploughed back into the soil, might have beneficial effects on soil organic matter content and structure, or that the frequent cultivation associated with arable cropping could allow for effective subsoiling.

In this context, the characteristics of the site, particularly in relation to soil type and climate, have a major effect on the most appropriate solution. On the wet, unstable soils in south Wales, it would be neither realistic nor desirable to consider crops other than long-term grassland. Regular or repeated surface cultivations have been seen to result in slumping of the soil, reduction in aggregate stability and retardation of earthworm population recovery, which is considered to be so central to the rehabilitation of structure on this particular soil type.[26] Thus relatively extensive use of grassland for grazing and for hay or silage production are the best options. In the Bryngwyn Farm Project, grass/clover swards were sown but nutrient inputs were restricted, since increased treading damage associated with higher productivity and heavier stocking rates would have negated the soil improvement programme which was based on the development of a biologically active soil.[26]

On the other hand, the clay soils on the restored sites at Butterwell and at Acklington in the north-east of England, in an area of lower rainfall, could be used to produce very satisfactory crops of autumn sown cereals, intensive grassland for grazing or silage, and oilseed rape, provided an effective drainage scheme had been installed. Table 9 shows that yields of wheat and grass on a newly restored site were only slightly lower than those at an adjacent unworked site of the same soil type and managed in the same way. In that locality, most farmers practise a mixed arable and grassland rotation of crops on unworked soils, and the inference from these results is that they should be able to continue to do so on land after restoration. Although earthworms are important, it is recognized that their populations are often low on frequently cultivated soils, and that in this situation some of their functions are fulfilled by the cultivation process.

One crop which was relatively unsuccessful was spring barley, because of the difficulty of preparing an adequate seedbed early enough in spring. This problem was encountered on both unworked and restored sites because of the characteristics of the soil type, but was most apparent on restored land. Cultivations in the autumn, on the other hand, were relatively easy. It was interesting to note that measurements of rooting depth and root mass were very similar for grasses and cereals and that, at this site, moisture was extracted from depth by both crops, inducing wetting and drying and the creation of root channels at depth which may lead to particle aggregation and the development of structure.

Use of fertilizer nutrients, especially nitrogen, not only stimulated the production of more above-ground biomass, but also resulted in the greater removal of water from the profile; so in this case there was no conflict of interest between short-term yield and longer-term soil development. At Butterwell, where different parts of the site had been subjected to completely different cropping regimes during the five year rehabilitation phase (*e.g.* continuous wheat or grass for five years or cereal/grass rotations), a uniform test crop of barley was grown on the site in year six. Table 10 shows that there was little or no evidence of residual effects of previous cropping.

The cultivation of green manure crops such as rye and oats/peas, together with the regular incorporation into the soil of the biomass produced, did not result in any significant improvement in soil organic matter status or soil structure on a site restored with previously stored soil. In fact, attempts to produce spring seedbeds were rarely successful and the yields of the spring-sown crops were often low. This, combined with the cloddiness resulting from attempts to produce tilth under marginally suitable conditions, meant that soil structure improvement was not achieved. Often crop root growth and moisture

Table 9 Comparison of winter wheat grain yield (t ha^{-1}) and grass herbage dry matter yield (t ha^{-1}) over four years at a site restored in 1981 and at an unmined control site

	Restored site	Control site
Wheat		
1982	10.2	12.5
1983	9.2	10.7
1984	8.0	11.5
1985	9.5	8.1
Grass		
1982	13.3	16.0
1983	10.1	12.0
1984	9.1	10.2
1985	11.8	12.0

Table 10 The effect of previous cropping history, from 1981–1985, on the yield of winter barley in 1986/87 on a restored opencast site (t ha^{-1})

Previous crop and management (1981–85)	Barley yield in 1987
Grass, nil fert. N, grazed	6.9
Grass, 200 kg N ha^{-1}, grazed	7.1
Grass, 400 kg N ha^{-1}, grazed	7.2
Winter wheat each year	7.4
Spring barley each year	7.0
Three years grass followed by wheat	6.8
SED	0.17

SED = Standard error of difference between means.

removal were considerably less good than in areas of the site with conventional crops of cereals or grass.

In terms of crop nutrient availability, most restored sites have pH values and phosphorus and potassium indices similar to those of surrounding unworked land, and the availability of these nutrients is not usually a major problem. Standard recommendations for application of these nutrients are normally quite satisfactory for restored opencast land. Nitrogen availability is influenced by the mining and restoration sequence, however, especially where soils have undergone a period of storage. After an initial flush of available nitrogen release following the spreading of stored soil and the reestablishment of aerobic conditions, the supply of available N is often considerably reduced. Whilst grassland experiments indicated that an extra 20–30 kg fertilizer N ha^{-1} could make up for the lower available soil N supply at the progressively restored Butterwell site, the deficit was greater at Acklington where stored soil had been used.

In an agricultural environment, the application of extra fertilizer N into a system which is already receiving substantial amounts (perhaps 250–300 kg ha^{-1}) is neither difficult nor excessively costly, so most farmers would be prepared to use that strategy if necessary to support yield. An alternative approach is to incorporate forage legumes such as red or white clover in grass seed mixtures because of their capacity to fix atmospheric nitrogen. Again, experiments have indicated that up to 200 kg N ha^{-1} can be fixed per year by white clover growing with ryegrass. This system, which depends upon the nodulation of clover roots by the nitrogen-fixing bacteria, was effective one year after soil replacement, without the inoculation of stored soil with any additional micro-organisms.

Another option is that favoured by the researchers in Wales, namely the use of organic manures from the animal production systems to maintain or increase soil nitrogen status. The attraction in their system is the encouragement of earthworm activity by organic manure addition, thus assisting in the development of soil structure as well as providing nutrients directly.

Thus in the very managed ecosystem which is modern agriculture, the cultivation of viable crops on land restored after opencast mining is not too difficult. Provided there is an effective drainage system installed, farmers can use accepted principles of crop cultivation and management, with slight modifications of existing practices, to grow economically viable crops. The more successful the crop, the better it is likely to be in terms of soil improvement.

Increasingly, there is a need to be able to cultivate restored areas for reasons other than agricultural output. For example, species-rich grasslands can provide interesting diverse ecosystems for many forms of wildlife, as well as being attractive in themselves. Experiments have been conducted to establish whether such habitats can be successfully created on restored land. Ecologically the main attraction of such habitats is the plant species diversity with perhaps 20–25 species m^{-2}, contrasting to agricultural swards with perhaps two or three major species. With care, seed mixtures typical of the area and situation can be selected and sown.[27] Their establishment and subsequent maintenance depends on numerous factors, not least the defoliation (cutting and grazing) strategy and the nutrient availability. Without question, species diversity is encouraged by allowing the crop to grow to a stage appropriate for hay cutting before defoliation, as this allows for seed setting and dispersal of certain species which depend upon that mechanism for their survival. It is also widely thought that species diversity will be encouraged by the low nitrogen status of restored sites. However, some experiments have shown that the N status may in fact be too low, allowing the nitrogen-fixing species to dominate to the detriment of other species, and restricting diversity.[28] Whilst it is possible to intervene by adding fertilizer N, this runs rather contrary to the objectives for some of these more natural ecosystems, which should eventually be self-sustaining. Ultimately, the challenge is to establish a system of organic matter and nutrient cycling which is at the correct level to maintain species diversity. This will depend upon the establishment and maintenance of a diverse soil biotic system which will necessarily be more complex than that required for commercial agriculture. This challenge remains to be fully addressed.

4 Conclusions

Coal-mining operations cause two types of land disturbance: derelict land covered with the spoil from underground mining, and surface-mined land which is excavated and later replaced. The problems faced in reclaiming the colliery

[27] R. Chapman and A. Younger, *Restor. Ecol.*, 1995, **3**, 39.

[28] R. Chapman and A. Younger, in *Grassland Management and Nature Conservation*, Occasional Symposium No. 28, British Grassland Society, Reading, UK, 1993, p. 248.

spoil sites are generally greater than for opencast sites, because little or no soil may be available on the former and the spoil material itself is not an ideal plant growth medium. On the other hand, the intended end use for reclaimed colliery spoil, either grassland or forestry, is generally less demanding than the varied agricultural cropping, which is often the aim of opencast restoration schemes.

The key problems to be addressed in reclaiming colliery spoil are, in order of importance, acid generation on some sites, compaction and related water-holding problems, and plant nutrient supply. All three problems can be reduced if soil is available as an amendment, and its use is recommended if at all possible. Small quantities of soil used as a cover layer can retard the acidification of the surface material, with as little as 15 cm depth of material being effective. Soil used either as a cover or incorporated into the spoil reduces the physical problems of compaction, poor water-holding and drainage, and also assists with nutrient supply. In the absence of soil, acidification has to be tackled primarily by locating the iron pyrites-rich material during site investigation, and then ensuring that it is buried deep within the regraded spoil heap during site preparation. If that is unsuccessful and acid generating material is near the surface of the regraded heap, then liming will be required to neutralize the acid produced. The requirement for fertilizers, particularly nitrogen and phosphorus, will be greater in the absence of soil. Finally, the poor physical conditions can only satisfactorily be improved in the absence of soil by regular recultivation and reseeding of the site.

All of the above relate to the reclamation of colliery spoil to agricultural grassland, but similar problems occur for reclamation to forestry. Soil is not generally used in such schemes and the problem of poor soil water availability, leading to physiological drought, is minimized by keeping ground cover around the trees to a minimum to avoid the competition effect. The problem of compaction cannot be tackled by frequent recultivation, so the solution is to plant tree species adapted to poor soil physical conditions.

Opencast coal-mining sites are restored either following storage of the overburden and soil, or by progressive techniques in which such storage is minimized. In either case the replaced soil is liable to physical damage during handling, so that the key problem is how to improve the soil structure. For the stored soils the amount of handling is greater and thus the potential for soil structural damage is greater. Ensuring good drainage is an important prerequisite for soil structure improvement; it is now recommended that subsurface drainage systems are installed as early as possible to maximize the benefits from them. Soil structure can also be improved by growing high yielding crops of either grass or cereals, as appropriate for the soil and climatic conditions prevailing. The benefits from such crops are derived from their water use, which helps to dry the soil, and their inputs of organic material, such as roots and other crop residues, to the soil. In all but the wettest areas the usual mixed farming crops can be used and they can be managed in the normal way, with the exception of spring cereals which have not been successful, because the poor soil physical conditions do not allow for good seedbed production at that time of year.

Stored soils require extra fertilizer or manure nitrogen, because following spreading and the return of aerobic conditions there is a rapid production of nitrate-N from ammonium-N, which will have accumulated during storage;

thereafter there is a shortfall in nitrogen release. Apart from that particular problem, fertilizer or manure applications can be at normal agricultural rates. In non-agricultural restoration schemes, fertility management is more difficult. This is particularly so where the intention is to recreate semi-natural species-rich grassland. Adjusting the nutrient status to maintain species diversity has proved to be difficult.

The problems associated with the reclamation of land disturbed by coal mining for subsequent use in agriculture and forestry are now well understood; the techniques developed to overcome them have proved largely successful. Where further work remains to be done is in the recreation of natural or semi-natural ecosystems on such land.

5 Acknowledgements

Much of the research reviewed here was carried out at the University of Newcastle upon Tyne. For the colliery spoil reclamation work, we would like to acknowledge financial support from the UK Department of the Environment (contracts: DGR 482/7 and 428/85) and from the EC (contracts: 288-77-1 and 328-79-1 ENV UK), and scientific and technical support from members of the Derelict Land Reclamation Research Unit. For the opencast restoration work we acknowledge funding from the former British Coal Opencast Executive, and scientific and technical assistance from the Northern Research Group.

Remediation of Lead-, Zinc- and Cadmium-contaminated Soils

MICHAEL LAMBERT, GARY PIERZYNSKI,
LARRY ERICKSON AND JERRY SCHNOOR

1 Introduction

Lead, Zinc and Cadmium Pollution

Non-ferrous heavy metals in the environment can be a major hazard to human health. Lead, zinc and cadmium pollution are the focus of this article, because these three metals are among the most common heavy metals found as contaminants in superfund sites,[1] and also because of the close association of these metals in the environment. Of particular concern is lead, owing to its widespread use by man. Lead has found numerous applications since the days of ancient Rome,[2] and zinc has been used at least as long.[3] Cadmium has a much more recent history of utilization, and has only been mined since the start of the 20th century.[4]

Lead is not a nutrient to plants or animals, but instead can cause great harm to organisms.[5] Lead pollution has been called the most widespread environmental health hazard in America,[6] and because of its widespread nature presents a more serious environmental and health hazard than other heavy metals.[7] In contrast, zinc has not often been thought of as being especially toxic and is, in fact, an essential nutrient.[8] However, the known deleterious effects of elevated environmental zinc from anthropogenic sources merits serious concern, especially for its phytotoxic effects.[9-11] Just as in the case of lead, cadmium is not a nutrient, but instead is known to be highly toxic to plants and animals.[12-14]

[1] J. McClean and B. Bledsoe, *Behavior of Metals in Soils*, Groundwater Issue EPA/540/S-92/018, United States Environmental Protection Agency, Washington, 1992.

[2] H. Waldron, in *Metals in the Environment*, ed. H. Waldron, Academic Press, New York, 1980, p. 155.

[3] H. Carus, in *Zinc, the Metal, its Alloys and Compounds*, ed. C. Matheson, American Chemical Society Monograph 142, Reinhold, New York, 1959, p. 1.

[4] J. Nriagu, *Nature*, 1979, **279**, 409.

[5] F. Haghiri, *J. of Environ. Qual.*, 1974, **3**, 180.

[6] W. Lindsay, *Chemical Equilibria in Soils*, Wiley-Interscience, New York, 1979.

[7] Q. Ma, T. Logan and S. Traina, *Environ. Sci. Technol.*, 1995, **29**, 1118.

[8] R. Khandekar, R. Raghunath and U. Mishra, *Sci. Total Environ.*, 1987, **66**, 185.

[9] A. Weatherly, P. Lake and S. Rogers, in *Zinc in the Environment*, ed. J. Nriagu, Wiley, New York, 1980, p. 337.

[10] L. Kiekens, in *Heavy Metals in Soils*, ed. B. Alloway, Wiley, New York, 1990, p. 261.

[11] G. Bryan and W. Langston, *Environ. Pollut.*, 1992, **76**, 89.

[12] S. Yasumura, D. Vartsky, K. Ellis and S. Cohn, in *Cadmium in the Environment*, ed. J. Nriagu, Wiley, New York, 1980, p. 12.

Natural Occurrence of Lead, Zinc and Cadmium Minerals

Lead, zinc and cadmium are all chalcophile elements, meaning that they have a tendency to form sulfide minerals.[15,16] Galena (PbS), one of the most common sulfide minerals, is also the most important lead ore, and sphalerite (ZnS) is the most important zinc ore.[17] Because galena and sphalerite ore deposits often form together,[18] lead and zinc pollution are commonly spatially associated. Cadmium, present in the lithosphere at an average content of $0.2\,mg\,kg^{-1}$, is not as abundant as lead or zinc, which have average lithospheric abundances of 16 and $80\,mg\,kg^{-1}$, respectively.[6] Because of its relative scarcity, cadmium is only mined where it substitutes for zinc in sphalerite.[19] Thus lead, zinc and cadmium pollution often occur together.

Weathering products of galena (secondary minerals formed by oxidation and/or hydrolysis) include phosphate minerals, *e.g.* pyromorphite $[Pb_5(PO_4)_3Cl]$, cerussite $(PbCO_3)$ and plumbogummite $[PbAl_3(PO_4)_2(OH)_5 \cdot H_2O]$, which are all fairly common in nature.[20] Common secondary zinc minerals that form from the oxidation of sphalerite include smithsonite $(ZnCO_3)$ as well as hopeite $[Zn_3(PO_4)_2 \cdot 4H_2O]$ and zinc pyromorphite $[Zn_5(PO_4)_3OH]$.[20,21] In comparison, the secondary carbonate mineral octavite $(CdCO_3)$ is relatively rare,[22] and cadmium phosphates are even less abundant, perhaps because of the low activity of cadmium in weathering environments.[20]

Anthropogenic Sources of Lead, Zinc and Cadmium Pollution

There are a number of ways to categorize the anthropogenic sources of environmental heavy metal pollution. Primary, secondary and tertiary metal contamination categories have been suggested for the sites of metal extraction and refining.[23] More recently, sources of lead pollution have been classified by originating activity (agricultural, industrial or urban activities),[7] and this classification can apply to all heavy metals.

The soil is a major sink for lead from human industrial activity,[24] and for zinc and cadmium as well. Because of the substitution of cadmium for calcium in apatite (a phosphorus ore), the application of phosphate fertilizer to farmland can

[13] B. Alloway, in *Heavy Metals in Soils*, ed. B. Alloway, Wiley, New York, 1990, p. 100.

[14] R. Chaney and J. Ryan, *Proceedings of the Conference on Criteria for Decision Finding in Soil: Valuation of Arsenic, Lead, and Cadmium in Contaminated Urban Soils*, Braunschweig, FRG, in press.

[15] V. Goldschmidt, *Geochemistry*, Oxford University Press, London, 1958.

[16] K. Krauskopf, *Introduction to Geochemistry*, McGraw-Hill, New York, 1979.

[17] L. Berry and B. Mason, *Mineralogy: Definitions, Descriptions, Determinations*, Freeman, San Francisco, 1959.

[18] J. Maynard, *Geochemistry of Sedimentary Ore Deposits*, Springer, New York, 1983.

[19] C. Klein and C. Hurlbut, *Manual of Mineralogy*, Wiley, New York, 1993.

[20] J. Nriagu, in *Phosphate Minerals*, ed. J. Nriagu and P. Moore, Springer, New York, 1984, p. 318.

[21] P. Barak and P. Helmke, in *Zinc in Soils and Plants*, ed. A. Robson, Kluwer, Boston, 1993, p. 1.

[22] D. Fassett, in *Metals in the Environment*, ed. H. Waldron, Academic Press, New York, 1980, p. 61.

[23] J. Moore and S. Luoma, *Environ. Sci. Technol.*, 1990, **24**, 1278.

[24] B. Davies, in *Lead in Soil: Issues and Guidelines*, ed. B. Davies and B. Wixson, Science Reviews Limited, Northwood, 1988, p. 65.

cause cadmium accumulation in the soil.[25-27] The application of sewage sludge as a soil amendment to farmland provides another avenue for the introduction of heavy metals such as lead, zinc and cadmium to farmland.[28,29] Industrial processes such as mining, refining and manufacturing can contaminate soil near the site of the activity, and can be a pollution problem long after the site has ceased operating.[30-33] Urban activities that contribute to heavy metal pollution include the use of leaded paint, once thought to be the prime source of lead poisoning in children, but now understood to be one of a number of possible sources.[34] Leaded fuel is still the greatest single source of anthropogenic lead to the environment, despite the increasing use of unleaded alternatives.[35]

2 Geochemical Considerations

Lead, Zinc and Cadmium in Soils

The oxidation states of lead, zinc and cadmium in soils of typical pH are discussed in the literature.[6] Although it is possible for Pb^{4+} to form in some highly oxidized soils, under most conditions lead exists as Pb^{2+}. Zinc and cadmium in soils occur exclusively as Zn^{2+} and Cd^{2+}, respectively. Pb^{2+}, Zn^{2+} and Cd^{2+} are therefore the labile forms of the metals in soils, available for uptake by organisms.

The formation of insoluble lead phosphates, including pyromorphite, acts as an important sink for lead in soils.[36-37] Because the average lead content in soils is about $10\,mg\,kg^{-1}$, and the average phosphorus soil content is about $600\,mg\,kg^{-1}$, it is likely that phosphorus acts as a control on lead.[6] By the same reasoning, phosphorus probably acts as a control for zinc ($50\,mg\,kg^{-1}$ average soil content) and cadmium ($0.06\,mg\,kg^{-1}$ average soil content) as well. Insoluble heavy metal phosphates, such as pyromorphite and pyromorphite analogues for zinc and cadmium, may be key controls on the mobility of these metals in the natural environment.

[25] R. Bramley, *New Zealand J. Agric. Res.*, 1990, **33**, 505.

[26] W. De Boo, *Toxicol. Environ. Chem.*, 1990, **27**, 55.

[27] Q. He, B. Singh, *Water, Air, Soil Pollut.*, 1994, **74**, 251.

[28] L. Walsh, D. Baker, T. Bates, F. Boswell, R. Chaney, L. Christensen, J. Davidson, R. Dowdy, B. Ellis, R. Ellis, G. Gerloff, P. Giordano, T. Hinsley, S. Hornick, L. King, M. Kirkham, W. Larson, C. Lue-Hing, S. Melsted, H. Motto, W. Norvell, A. Page, J. Ryan, R. Sharma, R. Singer, R. Singh, L. Sommers, M. Sumner, J. Taylor and J. Walker, *Application of Sewage Sludge to Cropland: Appraisal of Potential Hazards of the Heavy Metals to Plants and Animals*, Council for Agricultural Science and Technology Report No. 64, Iowa State University, Ames, IA, 1976.

[29] B. Alloway and A. Jackson, *Sci. Total Environ.*, 1991, **100**, 151.

[30] M. Buchauer, *Environ. Sci. Technol.*, 1973, **7**, 131.

[31] S. Bradley and J. Cox, *Sci. Total Environ.*, 1987, **65**, 135.

[32] J. Neuberger, M. Mulhall, M. Pomatto, J. Sheverbrush and R. Hassanein, *Sci. Total Environ.*, 1990, **94**, 261.

[33] J.-D. Wang, C.-S. Jang, Y.-H. Hwang and Z.-S. Chen, *Bull. Environ. Contam. Toxicol.*, 1992, **49**, 23.

[34] P. Body, G. Inglis, P. Dolan and D. Mulcahy, *Crit. Rev. Environ. Control*, 1991, **20**, 299.

[35] J. Jaworski, J. Nriagu, P. Denny, B. Hart, M. Lasheen, V. Subrahamanian and M. Wong, in *Lead, Mercury, and Arsenic in the Environment*, ed. T. Hutchinson and K. Meema, Wiley, Chichester, 1987, pp. 3–16.

[36] J. Nriagu, *Geochim. Cosmochim. Acta*, 1974, **38**, 887.

[37] J. Nriagu, *Inorg. Chem.*, 1972, **11**, 2499.

Heavy Metal Mineral Solubilities

Table 1 is a compilation of some lead, zinc and cadmium minerals that are found in soils.[6,7,21,37,38] This list is not complete, but for all three metals, phosphates comprise the most insoluble minerals. Pyromorphite $[Pb_5(PO_4)_3Cl]$ is thought to control lead activities in natural systems,[36] and it is reasonable to assume that zinc phosphates such as the mineral hopeite $[Zn_3(PO_4)_2 \cdot 4H_2O]$ may regulate zinc in ecosystems.[39] However, cadmium phosphates are very rare in nature, a fact that may be attributable to the low ionic activity of cadmium in soil environments.[20]

Phase and Solubility Diagrams for Lead Minerals. Figure 1 is a phase diagram showing the stability fields of common lead minerals in soils. It is immediately apparent from the large size of the pyromorphite stability field that pyromorphite is the most stable lead mineral under acidic to neutral conditions.[20] Minerals that are shown in Table 1 to be less soluble than pyromorphite (plumbogummite, hinsdalite and corkite) all require alkaline environments to form, as well as available aluminium or iron. Cerussite $(PbCO_3)$ has a small stability field, and anglesite $(PbSO_4)$ is not present.

Figure 2 shows the solubilities of common lead minerals in soils.[6] At all pH values, pyromorphite $[Pb_5(PO_4)_3Cl]$ is less soluble than other lead phosphates and cerussite $(PbCO_3)$. Under alkaline conditions, cerussite is less soluble than fluoropyromorphite $[Pb_5(PO_4)_3F]$.

It can be seen from these two Figures that the addition of phosphates to reduce bioavailable lead in soils by forming pyromorphite will sequester the lead in its most insoluble mineral form. In this way the lead will be sequestered *in situ*, and its uptake by living organisms prevented or reduced.

Phase and Solubility Diagrams for Zinc Minerals. A phase diagram for common zinc minerals in soils is shown as Figure 3.[20] Smithsonite $(ZnCO_3)$ has the largest stability field, followed by hopeite and related zinc phosphates. Of the zinc phosphates, zinc pyromorphite and spencerite are stable only under very alkaline conditions, and tarbuttite, uncommon in nature, has a very limited stability field. Therefore, under normal soil conditions, zinc will form smithsonite or hopeite. Note that the pyromorphite field in Figure 1 overlaps both the smithsonite and hopeite fields in Figure 3. Pyromorphite has a much lower solubility than either smithsonite or hopeite, and this may account for the fact that lead appears to out-compete zinc for available phosphorus in soils.[21,40]

Figure 4 shows the solubilities of zinc minerals in soil.[6] Hopeite is less soluble than smithsonite except at high pH, where phosphate is fixed by hydroxyapatite. According to this diagram, both hopeite and smithsonite are more soluble than soil-Zn, meaning that they would dissolve and release zinc in soils. However, the position of the soil-Zn in Figure 4 may shift for soils of different pH,[6] and

[38] S. Santillan-Medrano and J. Jurinak, *Soil Sci. Soc. Am. Proc.*, 1975, **39**, 851.

[39] J. Nriagu, *Geochim. Cosmochim. Acta*, 1973, **37**, 2357.

[40] J. Cotter-Howells and S. Caporn, *Appl. Geochem.*, 1996, **11**, 335.

Table 1 Solubility products ($\log K_{sp}$) at 25 °C of some lead, zinc and cadmium minerals (in order of decreasing solubility product)*

Mineral	Chemical formula	$\log K_{sp}$
Zinkosite	$ZnSO_4$	3.4
Anglesite	$PbSO_4$	−7.7
Smithsonite	$ZnCO_3$	−9.9
Octavite	$CdCO_3$	−11.6
Cerussite	$PbCO_3$	−12.8
Tarbuttite	$Zn_2(PO_4)OH$	−26.6
Scholzite	$Zn_2Ca(PO_4)_2 \cdot 2H_2O$	−34.1
Hopeite	$Zn_3(PO_4)_2 \cdot 4H_2O$	−35.3
Phosphophyllite	$Zn_2Fe^{2+}(PO_4)_2 \cdot 4H_2O$	−35.8
(cadmium phosphate)	$Cd_3(PO_4)_2$	−38.1
Spencerite	$Zn_4(PO_4)_2(OH)_2 \cdot 3H_2O$	−52.8
Zinc pyromorphite	$Zn_5(PO_4)_3OH$	−63.1
Fluoropyromorphite	$Pb_5(PO_4)_3F$	−76.8
Hydroxypyromorphite	$Pb_5(PO_4)_3OH$	−82.3
Chloropyromorphite	$Pb_5(PO_4)_3Cl$	−84.4
Hinsdalite	$PbAl_3(PO_4)(OH)_6SO_4$	−99.1
Plumbogummite	$PbAl_3(PO_4)_2(OH)_5 \cdot H_2O$	−99.3
Corkite	$PbFe_3^{3+}(PO_4)(OH)_6SO_4$	−112.6

*Modified from: Lindsay;[6] Santillan-Medrano and Jurinak;[38] and Nriagu.[20,37,44,47]

Figure 1 Stability fields of lead minerals in soils. Ionic activity constraints are: activity of SO_4^{2-} = activity of HCO_3^- = 10^{-3}; activity of Al^{3+} = 10^{-6}; and activity of Pb^{2+} = 10^{-6}. Minerals with similar compositions and stabilities, such as hinsdalite and corkite, occupy the same stability field. Chloropyromorphite [$Pb_5(PO_4)_3Cl$] has the largest stability field. (Adapted from Nriagu[20])

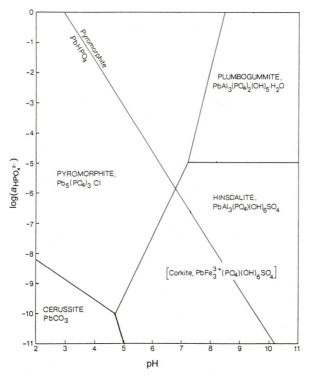

Figure 2 Solubilities of lead minerals in soils. Phosphate activity is fixed by strengite (FePO$_4$·2H$_2$O) and soil-Fe at low pH, by β-TCP [Ca$_3$(PO$_4$)$_2$] and soil-Ca at neutral pH values and by β-TCP and calcite at high pH. CO$_{2(g)}$ is 0.003 atm. Pyromorphite [Pb$_5$(PO$_4$)$_3$Cl] is the least soluble mineral at all pH values. The influence of hydroxyapatite [HA, Ca$_5$(PO$_4$)$_3$OH], tricalcium phosphate (TCP) and dicalcium phosphate (DCPD) as a control on phosphate is shown. (Adapted from Lindsay[6])

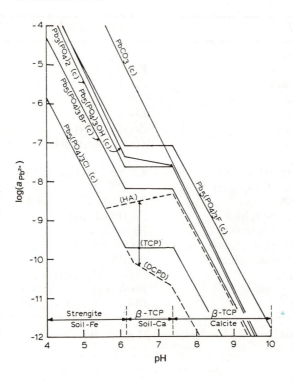

Figure 3 Stability fields of zinc minerals in soils. Ionic activity constraints are: activity of Fe^{2+} = 10^{-6}; activity of Ca^{2+} = activity of HCO$_3^-$ = 10^{-3}; and activity of Zn^{2+} = 10^{-6}. The zinc phosphate minerals hopeite, phosphophyllite and scholzite, with similar compositions and stabilities, occupy the same stability field. The relatively soluble zinc carbonate mineral smithsonite (ZnCO$_3$) has the largest stability field. (Adapted from Nriagu[20])

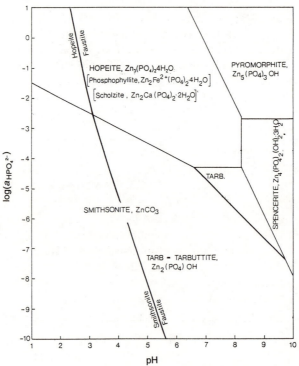

Figure 4 Solubilities of zinc minerals in soils. Phosphate activity is fixed by strengite ($FePO_4 \cdot 2H_2O$) and soil-Fe at low pH, and by DCPD ($CaHPO_4 \cdot 2H_2O$), β-TCP [$Ca_3(PO_4)_2$] or hydroxyapatite [HA, $Ca_5(PO_4)_3OH$] at higher pH, with Ca^{2+} fixed by soil-Ca or by calcite ($CaCO_3$) and $CO_{2(g)}$. The influence of hydroxyapatite (HA), β-TCP and DCPD on hopeite [$Zn_3(PO_4) \cdot 4H_2O$] solubility is shown in the diagram. Hopeite is less soluble than smithsonite ($ZnCO_3$), except at high pH when hydroxyapatite controls phosphate activity. At all pH values, hopeite and smithsonite are more soluble than soil-Zn. (Adapted from Lindsay[6])

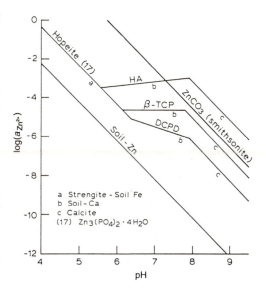

a Strengite - Soil Fe
b Soil - Ca
c Calcite
(17) $Zn_3(PO_4)_2 \cdot 4H_2O$

if hopeite were generally unstable in soils it would not be reported as a common weathering product of primary zinc ore.[20]

Solubility Diagram for Cadmium Minerals. Figure 5 shows the solubility of different cadmium minerals in soil.[6] In acidic to neutral soils, none of the cadmium minerals shown [an unnamed cadmium phosphate, $Cd_3(PO_4)_2$, and octavite, $CdCO_3$] is less soluble than soil-Cd. Octavite becomes stable in soils at high pH, although the use of TCP [$Ca_3(PO_4)_2$] to increase phosphate activity would make $Cd_3(PO_4)_2$ the stable phase. The wide range of low pH environments in which cadmium is not sequestered in a mineral phase may account for the observation that, under a given set of conditions, cadmium activity is always greater than that of lead.[38]

3 Soil Remediation

Types of Remediation

Soil remediation is a complex area with great need for additional research, not only in drafting clean-up guidelines, but also in simply defining what kinds and levels of contamination warrant remediation.[41] However, all types of remediation can be classified in one of two categories: (1) decontamination or (2) isolation and/or containment.[37] Some studies have been concerned primarily with phytoremediation (the use of green plants that accumulate heavy metals to decontaminate soil).[42,43] Phytoremediation decontaminates the site of pollution,

[41] S. Sheppard, C. Gaudet, M. Sheppard, P. Cureton and M. Wong, *Can. J. Soil Sci.*, 1992, **72**, 359.
[42] W. Berti and S. Cunningham, in *Trace Substances, Environment and Health*, ed. C. Cothern, Science Review, Northwood, 1994, p. 43.
[43] P. Kumar, V. Dushenkov, H. Motto and I. Raskin, *Environ. Sci. Technol.*, 1995, **29**, 1232.

Figure 5 Solubilities of cadmium phosphate and octavite (CdCO$_3$). Several different controls on phosphate are shown, including FePO$_4$·2H$_2$O (strengite) and Fe(OH)$_3$ at low pH, (soil-Fe) tricalcium phosphate (TCP), and soil-Ca at intermediate pH, and TCP, calcite and CO$_2$ at high pH. At low to neutral pH, the minerals are more soluble than soil-Cd. (Adapted from Lindsay[6])

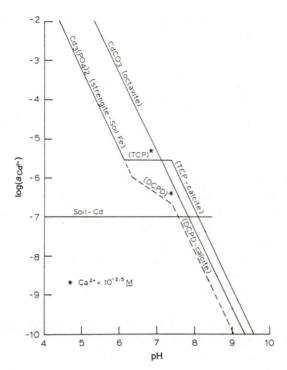

but leaves the problem of disposing of heavy metal-contaminated plant material. Isolation or containment involves *in situ* immobilization of the heavy metals in some insoluble form. One promising approach for *in situ* treatment of heavy metal pollution is the sequestration of heavy metals on-site in insoluble phosphate minerals such as pyromorphite.[21,36,44]

Phosphate Remediation of Lead-, Zinc- and Cadmium-polluted Soils

Much of the literature discussing phosphate remediation of non-ferrous contaminated soils deals with lead remediation.[7,45,46] This is no doubt due in part to the high public interest in lead poisoning, as well as to the dramatic reduction in lead solubility that can be brought about by pyromorphite mineralization.

One study[7] utilized column experiments in which ground natural phosphate rocks were mixed with Pb(NO$_3$)$_2$, precipitating lead phosphates and immobilizing an estimated 39–100% of the lead. The exact form of pyromorphite varied from chloropyromorphite [Pb$_5$(PO$_4$)$_3$Cl] to hydroxypyromorphite [Pb$_5$(PO$_4$)$_3$OH] to fluoropyromorphite [Pb$_5$(PO$_4$)$_3$F], depending on the available anions in the phosphate rocks. The solubility products (log K_{sp}) of these minerals range from

44 J. Nriagu, *Geochim. Cosmochim. Acta*, 1973, **37**, 367.
45 M. Rabinowitz, *Bull. Environ. Contam. Toxicol.*, 1993, **51**, 438.
46 M. Ruby, A. Davis and A. Nicholson, *Environ. Sci. Technol.*, 1994, **28**, 646.

-84.4 to -82.3 to -76.8 $\log K_{sp}$,[37,44,47] demonstrating the effectiveness of phosphate amendments in reducing the bioavailability of lead. Examination of lead-bearing soil with both lead and phosphate contamination from nearby industry shows that, over a period of 13 years, 46% of the original lead sulfide contamination was converted to lead phosphate (primarily pyromorphite) in what amounted to a long-term, uncontrolled field experiment.[46] A laboratory study reacted hydroxyapatite with lead-contaminated soil,[48] resulting in precipitated hydroxypyromorphite and a reduction in aqueous lead from $2273 \mu g l^{-1}$ to $36 \mu g l^{-1}$.

Phosphate remediation of lead- and zinc-contaminated soil by Na_2HPO_4 has been investigated in the laboratory,[40] where it was found that hydroxypyromorphite and an unnamed zinc phosphate had formed in a lead- and zinc-contaminated soil, and that a calcium-rich pyromorphite had formed in a calcareous soil contaminated by lead. It was concluded that whilst vegetative remediation techniques might be socially more acceptable than adding soluble phosphorus to soil, phosphorus amendment was shown to be more effective in reducing pollution. However, vegetative remediation can be used together with phosphate amendments to reduce both heavy metal bioavailability and the transport of heavy metals from the site of pollution. Although this study concluded that phosphate amendment can precipitate zinc phosphates as well as lead phosphates, there is controversy over how common a phenomenon zinc phosphate precipitation is in soils,[49] perhaps because of the belief of some workers that zinc phosphates such as hopeite are more soluble than soil-Zn.[6]

In another laboratory study, lead and cadmium remediation was attempted in smelter slag with the use of two different sources of phosphorus (KH_2PO_4 and apatite).[50] Lead remediation by the formation of pyromorphite was clearly evident, especially when the more soluble KH_2PO_4 was used. There was also some reduction in bioavailable cadmium, but not as obvious as that for lead (perhaps an indication of the lack of a suitable insoluble mineral phase under normal soil conditions). A laboratory study,[51] in which ammonium hydrogenphosphate $[(NH_4)_2HPO_4]$ was added to jars containing poplar cuttings, did not seem to influence cadmium uptake. It has been pointed out[38] that no comprehensive study has been done for soil phosphate as a factor controlling the solubility of cadmium.

The rate processes in soils are affected by diffusion and mass transfer. Further research is needed to understand fully the effects of phosphate amendments on soluble heavy metal concentrations in soils with high heavy metal content. The frequency of phosphate addition required to maintain a desired phosphate

[47] J. Nriagu, *Geochim. Cosmochim. Acta*, 1973, **37**, 1735.

[48] Q. Ma, S. Traina and T. Logan, *Environ. Sci. Technol.*, 1993, **27**, 1803.

[49] L. Shuman, in *Zinc in the Environment. Part I: Ecological Cycling*, ed. J. Nriagu, Wiley, New York, 1980, p. 39.

[50] M. Lambert, G. Pierzynski and G. Hettiarachchi, *HSRC WERC 11th Joint Conference on the Environment Abstracts Book, Albuquerque, NM, 1996*, p. 107; http://www.engg.ksu.edu/HSRC/Proceedings.html.

[51] D. Aoki, Master's thesis, University of Iowa, Iowa City, IA, 1992.

concentration in the aqueous phase depends upon the available sinks for phosphate.

Health Benefits of Phosphate Remediation

Owing to different minerals having different solubilities and rates of dissolution in the human gastrointestinal tract,[52] the mineralogy of lead-containing minerals is of great importance in determining the health risk to humans of lead contamination. The US Environmental Protection Agency has devised a mathematical model to estimate lead health risk, called the IEUBK (Integrated Exposure Uptake Biokinetic) model.[53] Uptake factors include such things as the total and bioavailable lead concentrations in the soil. Bioavailability of the lead is estimated in the biokinetic part of the calculations, which compares a particular lead-containing mineral or soil to lead ethanoate (lead acetate, assumed to be completely bioavailable).

Figure 6 shows an IEUBK model output for blood lead concentrations for children living in an area with $800 \, mg \, kg^{-1}$ soil-Pb (a suggested clean-up level for communities near lead smelters[54]). The EPA goal is for more than 95% of the children under six years old in a given area to have less than $10 \, \mu g \, dl^{-1}$ blood lead. Using a recent estimate[55] for the reduction in bioavailability of lead after phosphate remediation from 17.8% to 4.3%, the results are dramatic. Without phosphate amendment, 23,8% of the children in this example would exceed the blood lead guideline, while with phosphate amendment, only 10.6% of the children would fall into this category. In order to reach the EPA guideline for blood lead concentration in children, the IEUBK model can be used to show that without phosphate admendment the soil lead concentrations would have to be brought down to $415 \, mg \, kg^{-1}$, while with phosphate remediation the lead soil level would only have to be brought down to $600 \, mg \, kg^{-1}$. Thus, phosphate *in situ* remediation of lead-polluted soils can have immense health and practical benefits for the local population.

4 Summary

Lead, zinc and cadmium pollution often are associated, and are of growing environmental concern. This is especially true of lead and cadmium pollution as these elements, unlike zinc, have no biological function and therefore can only have harmful effects on living organisms when present in the soil in bioavailable forms. The three metals are often found together in natural ore deposits because

[52] R. Renner, *Environ. Sci. Technol.*, 1995, **29**, 256A.

[53] US Environmental Protection Agency, *Guidance Manual for the Integrated Exposure Uptake Biokinetic Model for Lead in Children*, Office of Solid Waste and Emergency Response, US Government Printing Office, Washington, 1994.

[54] US Environmental Protection Agency, *Proposed Plan: Residential Yard Soils, Oronogo-Duenweg Mining Belt Site, Jasper County, Missouri*, US Environmental Protection Agency, Kansas City, KS, 1996.

[55] T. Logan, S. Traina, J. Heneghan and J. Ryan, in *Proceedings of the Third International Conference on Biogeochemistry of Trace Elements, Paris*, 1995.

Figure 6 An Integrated Exposure Uptake Biokinetic (IEUBK) model output for a soil with 800 mg kg^{-1} soil lead. Two different levels of bioavailability are assumed: 17.8% (corresponding to no phosphate remediation) and 4.3% (corresponding to phosphate stabilization of soil lead). All other model parameters are default values. Phosphate remediation under these conditions results in a decrease of children with blood lead levels above EPA guidelines from 23.80% to 10.61% of the children.

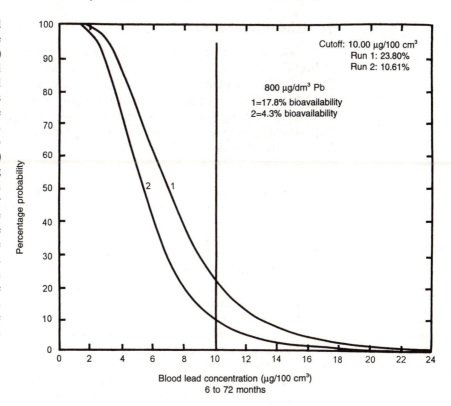

Cutoff: 10.00 µg/100 cm³
Run 1: 23.80%
Run 2: 10.61%

800 µg/dm³ Pb
1=17.8% bioavailability
2=4.3% bioavailability

Percentage probability

Blood lead concentration (µg/100 cm³)
6 to 72 months

of their frequent occurrence as sulfides, and so many polluted sites suffer from contamination from all three metals. *In situ* isolation of heavy metal pollution has the advantage of not producing a contaminated by-product that must be safely disposed of, and the sequestration of lead, zinc and cadmium in insoluble minerals such as phosphates would appear to be an ideal *in situ* remediation strategy. This has been shown to be very effective for the formation of lead phosphates, and there is some indication that zinc phosphate formation may also be effective. The case with cadmium is less clear, and would benefit from more study.

The health and practical benefits of phosphate remediation are made clear by the US Environmental Protection Agency IEUBK model. For a hypothetical example of an area with 800 mg kg^{-1} soil lead, phosphate remediation to precipitate pyromorphite reduces the estimated percentage of local children under the age of six years with blood lead levels above 10 µg dl^{-1} from 23.80% to 10.61%. Phosphate amendment can also be shown to allow this hypothetical area to meet the EPA guideline of less than 5% of children under six years of age having no more than 10 µg dl^{-1} blood lead by reducing soil lead to 600 mg kg^{-1}, instead of the 415 mg kg^{-1} that would otherwise be required.

5   Acknowledgements

The work presented here was partially supported by the US Environmental Protection Agency (EPA) under the assistance agreement 819653 for the Great Plains–Rocky Mountain Hazardous Substance Research Center. It has not been presented to the EPA for peer review and therefore may not reflect the views of the agency and no official endorsement should be inferred. The Kansas Agricultural Experiment Station, and the Center for Hazardous Substance Research also provided partial funding.

Risk Assessment and Management Strategies

GWYN GRIFFITHS AND STEVE SMITH

1   The Background to Land Reclamation in Wales

The Welsh Development Agency (WDA) was created in 1976 and charged with the task of helping to regenerate the Welsh economy. The means by which it addresses this task are as follows:

1. Providing sites and premises for industry
2. Providing support to industry
3. Attracting inward investment
4. Urban regeneration
5. Improving the Welsh environment

The dominant feature of the Agency's work in improving the Welsh environment has been the reclamation of derelict land, carrying on the task started by the Welsh Office in the 1960s largely in response to the Aberfan disaster.[1] The Agency's reclamation work also plays an important role in support of the first, third and fourth activities listed above.

In carrying out its land reclamation activities the Agency has always worked in close partnership with Local Authorities who receive 100% grant aid for the acquisition of derelict land, and then the design, supervision and implementation of land reclamation works. The 1982 Derelict Land Act increased the Agency's powers by enabling it to offer grants to the private sector for derelict land reclamation, but this grant is limited to 80% of the net loss involved in the activity and is therefore not particularly attractive.

The Agency is responsible for the management, control and funding of the reclamation process throughout the whole of Wales. The key tool in this process is a five-year rolling programme of projects. Such an approach was developed very early in the days of land reclamation in Wales to counteract the many impediments which can arise in the progress of any land reclamation scheme, be it land acquisition, design complications or, in the later stages of the process, the vagaries of weather. The flexibility of the rolling programme ensures that projects can be brought forward from a larger pool of selected projects to prepare a

---

[1] Report of the tribunal appointed to enquire into the disaster at Aberfan on 21 October 1966, HMSO, London, 1967.

development site for a specific need or to make good any budget slack which might arise as a result of unpredictable delays. The programme can also cater for the opportunist addition of one-off new projects occurring as a result of sudden closure, or newly identified hazard.

The Agency's tasks in managing the reclamation process in Wales are:

1. A constant process of review and assessment of existing and new dereliction in dialogue with Local Authorities and other interested bodies and potential developers
2. Systematic assessing and prioritizing of projects to maintain an agreed and approved rolling programme of projects for implementation
3. Management of the technical and financial content of particular projects and management of the overall rolling programme, including project and budget phasing
4. Encouraging innovation and supporting, promoting and participating in research work
5. Monitoring of reclaimed land

The last process is important to ensure that all after-value arising on sites is returned in a timely manner to help fund future activities. The monitoring process also informs future reclamation decisions in two ways. The first of these is to ensure that the physical reclamation techniques being applied are durable and effective, and the second is to ensure that the designation of land uses is successful to the extent that the intended uses are being achieved. The importance of checking the durability of reclamation works cannot be over emphasized because, although reclamation tends to use many existing techniques, they are often used at a scale and in a context not previously experienced. Furthermore, reclaimed sites need to survive in perpetuity with minimum maintenance.

The single most significant stimulus to land reclamation in Wales was undoubtedly the Aberfan disaster in October 1966. Prior to that date, only three derelict sites had been reclaimed. These extended to approximately 100 acres and had cost £28 000 to reclaim. The tragic loss of 144 people as a result of the colliery spoil-tip slide at Aberfan resulted in Government taking immediate action by setting up the Welsh Office Derelict Land Unit, it being charged with bringing about the reclamation of dereliction in Wales. Since 1966, 1200 sites have been reclaimed in Wales at a cost of approximately £350 million. These sites extend to a total of 10 000 hectares. The early practitioners in the field of land reclamation in Wales, driven by a deep-seated urgency to make rapid progress, sought wherever possible to achieve simple cost-effective solutions, so maximizing the amount of land reclaimed for a limited resource. Although the early reclamation may have been a little short in finesse in terms of revegetation, many of those initial shortfalls have been redressed by nature working unaided over time. This in itself provided a valuable lesson as to the advisability of relying upon nature as much as possible. With experience, however, greater resources are now directed towards the revegetation of sites.

Inclusion of the land reclamation role within the remit of the WDA has brought the significant advantage that the land reclamation team is working within an organization which owns, develops and disposes of reclaimed land in the market place. It is also within an organization responsible for attracting inward investment. These pressures have helped in focusing the team's activities on creating land suitable for development, by meeting the requirements of the developers or investors whilst acknowledging the need to create a safe and attractive environment for existing communities and potential inward investors to invest, live and work in.

Contrary to the generally held perception, only about 50% of Welsh dereliction is directly related to coal mining. The remaining 50% arises from a wide range of origins such as iron and steelworks, gas works, power stations, docks, railways, chemical tips, military dereliction, slate, hard rock quarrying, a wide range of factories and the 750 non-ferrous metal mining sites spread throughout Wales. It is not surprising, however, that the dominant land reclamation activity in the first decade after Aberfan was heavily orientated towards colliery and coal tip sites. The basic order of priorities used for tackling projects has always been that of firstly removing hazards, secondly reclaiming sites with development potential and thirdly achieving environmental improvements. Few sites fall into one clear category and many projects achieve all three outcomes.

Whilst one can never be complacent that tip stability issues have been totally dealt with, there has undoubtedly been a diminution in the number of sites remaining which pose a physical threat. However, as that workload has fallen off, the burden posed by contaminated sites has grown, fuelled by the identification and availability of such sites, coupled with a demand for their treatment and the derivation of acceptable treatment techniques.

Despite sites with development potential having been given very high priority, the availability of such sites has not ceased, but has continued to be refuelled with new closures throughout Wales in the early 1980s and with the demise of the coal industry in South Wales in the early 1990s. The magnitude and setting of many of the large eyesores in Wales has been justification enough to tackle them, but, of course, the removal of such eyesores usually brings with it major environmental benefits in terms of the prevention of erosion, dust generation and flooding, and the removal of risks associated with derelict buildings, shafts, structures and debris.

The Reclamation Team in the Agency is privileged to sit at the centre of all of the specialists working in land reclamation within Wales. It is therefore able to act as a link between those specialists and other public bodies and specialists throughout the UK and abroad. Seeing all of the project designs throughout Wales, it is able to act as a sorting house for promising or proven innovation and, where necessary, can prevent the repetition of seemingly promising innovations which have proved to be less than wholly effective.

Land reclamation is unique within the field of construction activity, in that the design process actually defines the end product. This is obviously at variance with the normal process where the first step is definition of the product, be it a road from A to B or a structure to contain a defined activity such as living, manufacture or some service activity. This aforementioned unique quality of land reclamation, whilst giving scope for the flare and ingenuity of the designer, also enables the design process to be one of iteration between what is the most desirable land use in simple planning terms and what is the most cost effective use of the site, taking account of the cost and achievability of the standard of reclamation required for the particular end use.

Because much of the Agency's early work was focused on colliery spoil sites, most of the early research was geared to the successful revegetation of these sites. It is salutary to realize that the early advice given by the Agricultural Development and Advisory Service (ADAS) regarding the revegetation of South Wales upland colliery spoil sites still holds good. This advice indicated that for fast germination and stabilization of the colliery spoil surfaces, normal agricultural grass species should be used in the knowledge that, on the shale surfaces virtually devoid of nutrient, these grasses would regress in time and be replaced progressively by the invasion of local indigenous species. There is now ample evidence that, despite the depredations of the ubiquitous Welsh sheep, this process is progressing very well. ADAS also recommended the use of broiler house litter at 12 tonnes per hectare as the initial nutrient needed to start this process. This also disposed of what was becoming a major waste product embarrassment in the agricultural industry and this was a good example of innovative recycling.

On careful reflection, it is true to say that both Welsh and other land reclamation activities in other areas relied heavily for their innovation upon the full-scale testing out of new ideas. This was not done as some costly, frivolous, full-scale experiment, but on the basis that it would certainly achieve 90% of the anticipated effect, and that a minimum 90% effect was well worth achieving if it brought with it the prospect of an overall improvement across the whole land reclamation process.

2 Emergence of Contaminated Land as a Reclamation Issue

Although it was as early as the 1970s that the WDA and its predecessor funded research by Liverpool University into the assessment, prioritization and reclamation of lead wastes sites throughout Wales, it was not until the late 1980s that a major element of the Agency's spending on land reclamation was directed to the more difficult contaminated sites. To date, approximately 50 sites have been reclaimed where some of the works were geared towards combating some level of contamination, and a further 40 sites of a total of 250 are currently being designed or are in progress. Whilst this figure suggests that only 6% of sites in Wales have been adjudged to merit treatment of their contamination, this should be viewed in the context of Welsh industry which, unlike that in, for instance, the Scottish Central Lowlands, the West Midlands or Lancashire, has been heavily orientated to the production and sale of prime products such as coal, steel ingots and rails, with little secondary industry and its more complicated and contaminated by-products. This is not to say that Wales does not have its fair share of contaminated sites, such as town gas works, coking plants, lead wastes sites, smelting sites and more contaminating industries, but that many of these sites are still in active use and are not therefore available for reclamation.

The move to reclaim contaminated sites was usually spurred by a particular local or national initiative. The most dramatic of these was the Lower Swansea

Valley. Action to ameliorate this immense industrial scar grew from some early academic and physical work by Swansea University in the 1960s, and the far-sighted desire of the Swansea City Fathers to see a major problem removed and opportunities created. This site, which extends to approximately 220 hectares, had been totally despoiled by a mixture of coal mining, iron and steel works, copper, lead and zinc smelting, and had been left virtually devoid of vegetation and covered with sprawling masses of slag and wastes of every description. The early work focused on the simple regularizing of the surface features, often by hand, and planting of large areas of trees. The more disturbed and derelict floor of the valley was at this stage left untouched because it still contained a few scattered relics of active industry. Between the mid-1970s and mid-1980s, Swansea City Council drove forward a comprehensive programme of reclamation funded by the Welsh Development Agency. The design philosophy consisted of, where possible, channelling surface and ground water through the site and away from the contaminants, reprofiling, redistributing and concentrating the contaminated wastes, and capping them over with locally won material; the excavated lake at the centre of what subsequently became the Enterprise Park was a major source of good quality clays. Although if undertaken today this major programme of works might be carried out with greater rigour, nothing should be allowed to detract from the magnificent achievements on this site in terms of reducing the availability of heavy metals, particularly into the water regime, and the adequacy and satisfactory nature of the site for the purposes to which it has been put, coupled with the visual transformation of what had been a uniquely appalling vista at the entrance to the city.

This site does perhaps best illustrate those truisms that if one wants to reclaim to standards appropriate to the year 2000, one should not start work until the year 2000 and, secondly, that whilst one may suggest that one should also seek the perfect solution, perfection is not a sustainable concept in that, almost by definition, it carries with it a near infinite time penalty and a cost also approaching infinity.

The early research work into the reclamation of lead waste sites, which was carried out by Liverpool University under the guidance of Professor Bradshaw,[2] leant heavily on the use of resistant species, in particular a native red fescue. Whilst the concept has merits and does undoubtedly offer scope to revegetate lead waste sites at low cost, experience at Parc Mine, Llanrwst, showed that steep sided, heavily contaminated, man-made spoil heaps in areas of high rainfall certainly needed major engineering works prior to revegetation to ensure control of erosion. At Parc Mine the tip was completely regraded, covered with relatively inert imported subsoil, drained and then revegetated. Figures 1 and 2 illustrate the dramatic improvements at the site. After almost 20 years these works are still proving to have been successful, particularly in preventing the erosion of large volumes of heavily contaminated lead waste into the adjoining water course where it was, in turn, affecting both the local water regime and the River Conwy and its flood plain further downstream.

At Cwmsymlog near Aberystwyth a similar technique was applied, but the

---

[2] A. D. Bradshaw and M. J. Chadwick, *The Restoration of Land: the Ecology and Reclamation of Derelict and Degraded Land*, Blackwell, Oxford, 1980.

Figure 1 Parc Lead Mine
wastes exposed to wind
and water erosion

Figure 2 The revegetated
Parc Mine site

prime objective there was to control heavily contaminated dust which, as a result of a freak funnelling effect of the topography, resulted in house dust in local residences containing up to 4% lead. Whilst the capping and grassing of the tip reduced movement of contamination into the water course and prevented wind blow of metal-rich dust, it was not possible to address the problem of high zinc levels in the mine drainage waters. These continue to damage the streams in the area.

Following the relative success of the reclamation works at Parc Mine and Cwmsymlog, it has become standard working practice to enhance the capping of the lead waste tips with a plastic liner or break layer. This obviously has the dual benefit of preventing capillary action of soluble contaminants upwards into the capping layer and vegetation, and prevents the downward movement of rainwater which would, in turn, generate a leachate out of the tip. Schemes which have benefited from this approach include Minera at Wrexham, Y Fan near Llanidloes and Goginan near Aberystwyth.

Through the 1970s and 1980s the Agency had attempted to keep pace with the general thinking in terms of the remediation of contaminated sites, but it was not until the prospect of the Section 143 Registers[3] was viewed through the Surveyor's perhaps distorted magnifying glass that developers and their professional advisors started asking truly challenging questions as to what the reclamation of a site had entailed and what it meant both in terms of the current site condition and any likely future liabilities. This event coincided quite closely with the Agency's programme of disposal of much of its industrial land and premises. The number of transactions in which the Agency was engaged, coupled with the new level of detail being required by purchasers, meant that the Agency professional staff were heavily involved in researching the history of many sites. The quality of the fruits of these searches depended on:

1. The searcher's personal knowledge of the history of the project
2. His/her knowledge of those who had been involved in the design and implementation of the project
3. The whereabouts of those people or any relevant documentation

Even when lucky enough to be blessed with these basic pieces of knowledge, however, the findings were relatively limited.

The two pieces of documentation that could usually be found, even after a long intervening period since reclamation, were firstly a site investigation survey of the site, and secondly a set of contract documents and drawings setting out the intended works. It was perhaps inevitable that the audit trial had two important missing links. First, a rational explanation as to why the particular works had been implemented in the light of what the site investigation had revealed. The second omission was a true and accurate record of what had actually been found during the currency of the works and what had finally been done. As is often the case, the 'as-built' records of the projects were carried out very much as an afterthought, and frequently tended to focus on such things as changed fence lines or gate details rather than on the substance of the construction of the site.

3 The WDA Manual

In the light of this sobering experience it was quickly concluded within the Agency that a protocol had to be developed which ensured that all sites

---

[3] *The Environmental Protection Act* 1990, HMSO, London (Section 143 now repealed).

completed were accompanied by clearly documented site investigation, a clear explanation of the thinking behind the type of remediation adopted, a contract document and drawings as ever, but then a clear and concise record of what was found during the project, what was actually done on site and, where necessary, any validation measures carried out on completion of the project or which would be required in advance of any particular development.

Whilst the Agency has always resisted the request for any form of guarantee, it is accepted that as a responsible custodian of the public purse it must be prepared to show accurately and clearly what was found in a site, why a particular course of action was chosen, and what action was actually taken to deal with the site. Beyond that it is still incumbent upon a developer to accept a normal level of developers' risk and to satisfy himself that the site is appropriate for his particular form of development and use.

The development of a robust system for recording the investigation, design, implementation and validation stages of a project had to recognize the implications of technical, legal and management issues in a rapidly advancing field. Not only was it essential to deal with developer uncertainty, it was also vital to address the wider impact of public perception on environmental quality. South Wales, in common with many areas of declining traditional industry, was recognizing a by-product of the process of decline: that which allowed the local communities to once again enjoy clean air and green fields around and within their communities. Hence, their expectations for the clean-up of sites is very high.

The fundamental objective of remediation must be to ensure that hazards are treated to the extent that environmental liabilities and risks to users or occupiers are reduced to an acceptable level. UK policy and guidance has been developed on the basis of a 'suitable for use' approach. Consequently, the acceptable level of risk will vary in relation to the after use proposed for a particular site and the susceptibility of the local environment.

Determination of this level of risk did not have the benefit of a comprehensive guidance document in the early 1990s, but, in order to satisfy developers, it was clearly desirable to demonstrate that best practice techniques had been adopted. Against this background, the Agency embarked on the development of a management system which would achieve a consistent and professional approach by ensuring that:

1. Appropriate technical procedures are used to identify, investigate, assess and, where necessary, remediate contaminated land
2. Appropriate management procedures are used to confirm the validity and effectiveness of technical measures

Hence, it would be possible to demonstrate that a high standard of remediation has been achieved and the site is capable of supporting its intended use. The interrelationship between these two elements is a fundamental principle of successful remediation.

The management system is set out in a comprehensive guidance document, the WDA Manual on the Remediation of Contaminated Land,[4] which has been adopted on all remediation projects funded by the Agency. It requires project managers to adopt procedures aimed at ensuring that all contaminated sites follow a systematic, rigorous and fully documented process of investigation and assessment. The requirements of the Manual have recognized the site-specific nature of contamination, which prevents the application of prescriptive solutions for the treatment of a site. Each site will require a comprehensive project brief to be prepared in order to define the objectives of the project and the scope of the work required. It must be recognized, however, that some flexibility in the design process is advisable to cater for periodic reviews of the objectives as more information becomes available.

The appointment of highly experienced and appropriately qualified specialist personnel is vital to the success of the project. These specialists are required to be familiar with and understand the policy and legal context of their duties, the interaction between a range of disciplines (*e.g.* chemists, hydrogeologists and civil engineers) and the use of technically advanced methods for investigation and assessment. Inappropriate procurement decisions at the outset of a project can have significant adverse implications on management effort and resources.

From the client's perspective, it is essential to ensure that the responsibility of each party involved in the project, *i.e.* main consultant, sub-consultants, investigation and remediation contractors, is fully defined and that advice being given is truly independent. Conventional procurement will normally ensure that this is the case. Procurement of remediation on a 'design and build' basis will still require the client to retain independent management advice for the project.

The main consultant must also have strong management skills to provide leadership of a multi-disciplinary team. Contaminated land projects have significant potential for conflicting factors and requirements to arise between each of the disciplines. The lead consultant must be capable of providing a valid assessment and firm recommendations based on all of these competing factors. Furthermore, he/she must be able to provide the client with a satisfactory professional indemnity insurance policy, which will allow the client to obtain redress in the event of any subsequent failure to meet the project's objectives arising from negligence. The insurance industry is able to provide project-specific policies designed for the particular circumstances of the site. Such policies may be of benefit to the client in ensuring that financial risks specific to the site are covered.

The overall objectives can only be achieved by adopting a risk management framework for all stages of a project. Risk management provides a systematic, objective and transparent basis for the management of contaminated land. The principal advantages are:

1. Proper characterization and evaluation of the site
2. Selection of appropriate remedial strategies
3. Effective control or reduction of defined risks
4. Effective technical and financial control of the project

Risk management is defined as the 'process whereby decisions are made to accept a known or assessed risk and/or the implementation of actions to reduce

---

[4] *The WDA Manual on the Remediation of Contaminated Land*, Welsh Development Agency, Cardiff, 1994.

the consequences or probabilities of occurrence'.[5] The two main elements of the management process are risk assessment and risk reduction. Within each of these two headings are a number of inter-related and iterative stages which require a systematic approach to the identification and assessment of the hazard, leading to the estimation and evaluation of risk and, lastly, definition of the measures required for risk control.

A number of definitions, from various sources, are in use for these terms and only in recent years has the UK moved towards standardizing their meaning. Guidance being issued by the Department of Environment will lead to a common acceptance of the terminology. The principal objective of risk assessment will be to determine the probability of potentially adverse effects on human health or impairment of the environment and the magnitude of the consequences. Risk management should not be regarded as a discrete step in the remediation process. The definition clearly conveys its impact on each stage of a project, from site investigation to validation. Figure 3 illustrates the relationships between the main components of the process of remediation and the stages of risk management.

The initial step for any project, that of gathering data specific to the site, must provide an effective contribution to the management process. Consequently, the objectives and the strategy to be adopted for a site investigation have to be defined very clearly. There are numerous objectives to be considered in determining the site investigation strategy[6] and each makes an important contribution to the project. A phased approach is essential in order to make informed decisions and apply the data correctly. The strategy should seek to ensure that data are representative of the site conditions, but it is uneconomic and impractical to analyse for an infinite number of contaminants. The phased approach will ensure that characterization of the site recognizes the likely hazards to be encountered and the potential pathways and targets both within and external to the site.

The procurement of the investigation services must consider quality control issues for both on-site works and laboratory analysis. The nature of the contaminants likely to be encountered dictates the precautions to be taken in retrieving samples (*e.g.* Are they hazardous to the health of operatives? Are the substances volatile?) and preparing them for transport to the laboratory. 'Chain of custody' arrangements must be established to ensure that quality standards are maintained, and that loss, deterioration or mixing of samples does not occur. The laboratory must have adequate experience of the required analysis and techniques should be standardized. Approval under accreditation schemes for analytical services, *e.g.* National Measurement Accreditation Service (NAMAS) and CONTEST, should be offered by the laboratory.

The concentrations of contaminants identified in the site are used to determine whether a hazard exists (with reference to relevant standards), whether the risks associated with this hazard are acceptable and, if not, what remedial action is required. Definition of which reference standards to apply is a major issue in the risk assessment procedure. These could be generic soil quality criteria, generic

---

[5] Royal Society, *Risk: Analysis, Perception and Management*, Royal Society, London, 1992.

[6] J. Petts, in *Proceedings of a Conference on Site Investigation for Contaminated Sites*, IBC Technical Services, London, 1993, p. 3.

Figure 3 Relationship between risk management and stages of a project. (Taken from the Welsh Development Agency, 1994)

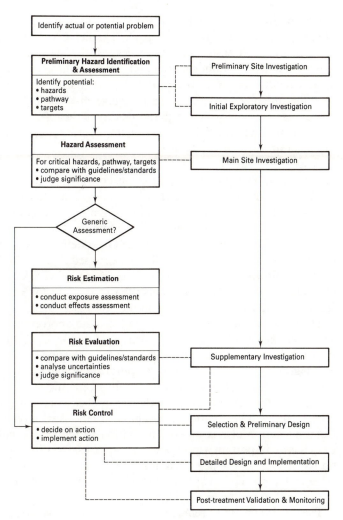

standards for other media (*e.g.* air or water) or other suitable standards, (*e.g.* Occupational Exposure Limits). Whatever generic standards are adopted, it is incumbent upon the specialist risk assessment personnel to recognize the basis on which the standards were derived and any inherent safety factors already accounted for.

Such qualitative risk estimations may be sufficient for many projects, but there will be instances when site-specific risk estimation leading to quantified estimates of risk is required. Typical cases where such an approach is justified are where generic standards are not available, complex problems prevail at a site (*e.g.* level of exposure is severe) or where local background conditions show high concentrations of contaminants. It is interesting to note that low levels of contaminant concentration may still constitute a hazard due to local circumstances and that high concentrations may be acceptable where no pathway exists to any sensitive target.

Site-specific risk estimation will require the use of models to describe the hazard/pathway/target relationship. Data gathered from the site investigation on contaminant concentration, properties of the pathway media (geological, hydrogeological or atmospheric) and potential targets will be utilized in the chosen model to define the probability of 'harm' occurring to the targets. Assumptions made in the estimation stage must be fully documented and a robust case presented to confirm the validity of the reference standards adopted. This will provide essential information for securing acceptance of the project by the client and regulatory bodies.

Where the assessment indicates that the risks associated with a contaminant are unacceptably high, consideration must be given to implementing risk control or reduction measures. A range of options will usually be available, but it is important to consider what standard of remediation is required prior to the preparation of a remedial strategy. The Agency's Manual adopted the concept of 'Contamination Related Objectives' (CRO) for defining the site-specific remedial standards to be achieved. The CRO is expressed as a residual concentration of contaminant which is acceptable for a specified level of risk. They may be generic guideline values, generic values adapted to account for site-specific factors or specifically derived values based on the risk assessment.

The CRO will be used for selecting the most appropriate remedial strategy. Their derivation will be fully recorded in a Final Investigation and Assessment Report. This report is an integral part of the reporting structure specified in the Manual, as shown in Figure 4. The reports are primarily aimed at informing the client and providing a discussion document for the various project approval stages. In addition, they provide the comprehensive record of the remediation process and will be a valuable tool in negotiations with regulatory authorities. Communication of risk-based information to regulators and other interested parties is an important aspect of managing risks. Decisions will need to be made on issues such as the degree and nature of the involvement of each party, the concern and perception of risks (a very real issue for local communities) and how the information is to be presented.

The development of a remedial strategy will need to take account of policy and administrative factors as well as technical factors. Land use is a major factor relative to the acceptable level of risk and the CRO must be consistent with the proposed use. As definition of the strategy progresses, or new data become available, it may be necessary to review the CRO or the use of the site in order to achieve a cost effective remediation. One feature of this concept, however, is the opportunity to adopt different objectives for different parts of the same site, dependent upon after-uses and targets at risk. Thus, original intentions for the site may still be achieved, albeit to a lesser extent or in a different form or layout.

This process of review is an integral part of the design and management procedures. The objectives and constraints should be prioritized to identify those on which design effort is to be focused and to overcome conflicts. This requires the design team, in conjunction with regulators and the client, to identify those objectives which must be achieved (*e.g.* to meet statutory requirements) and those which can be 'negotiated' as a trade off with other objectives (*e.g.* land use, time and cost criteria). The definition of priorities will be guided by the findings of the risk assessment to ensure that major hazards present on a site are always addressed.

Progress through the design stage will be an iterative process of selecting potentially feasible techniques (or combination of techniques), developing appropriate strategies and evaluating each strategy to determine its relative merits. The preferred strategy will be evaluated against the final selection criteria defined during the design and assessment process. These criteria should reflect both long-term performance and short-term operational aspects of the strategy. An example of final selection criteria is given in Table 1.

Implementation of the strategy will need to recognize the importance of technical, planning, supervision and management issues. These issues should be addressed in a formal management plan prepared during the detailed design/contract documentation stage of the project. The plan will include a clear indication of the objectives of the scheme, the options (and preferences) for contract procurement, the resource implications and respective responsibilities of each party. Other issues of equal importance are identified in Table 2. The plan will provide an essential reference point for subsequent monitoring and evaluation of the project, in addition to satisfying third parties (*e.g.* regulators) that satisfactory standards of quality control will be applied to the project.

Decisions on contract procurement will need to take account of the expertise available, in both consulting and contracting, to implement the preferred strategy. In most cases, conventional procurement of contracting services based on a fully detailed and specified remedial scheme will be appropriate. However, a 'design and build' procurement route will be necessary if the strategy requires a specialized technique or equipment which is only available in a small number of companies. This latter route can have the benefit of simplifying management requirements and apportionment of responsibility for the project. These can arise as a result of only one organization being responsible for all aspects of the work, including short- and long-term performance of the remedial treatment and quality control. Nevertheless, it is advisable for the client to retain an independent adviser to ensure that specified remedial standards are being achieved and full records of the site works are maintained.

Monitoring of the works (defined as the regular assessment of progress, performance and quality) will be the subject of a pre-defined Monitoring Plan in which responsibilities, objectives, acceptable sampling and analytical techniques and continency plans are established. The contractor will typically be responsible for implementing the monitoring programme but the client's independent design specialist, and in some cases the regulatory bodies, will need to have access to the data or the site for the purpose of ensuring compliance with the project objectives. Monitoring may extend beyond the completion of physical works, for example where containment has been adopted, to confirm the effectiveness of remedial measures.

The monitoring programme will identify the appropriate point at which validation (defined as confirmation of the project objectives being attained) is to be undertaken. The validation process is usually a one-off procedure and will result in an acknowledgement by the client that the contractor has discharged his contractual obligations in full. Validation should be undertaken within a reasonable timescale following completion of the works and it may be beneficial

Figure 4 Phasing and reporting structure. (Taken from the Welsh Development Agency, 1994)

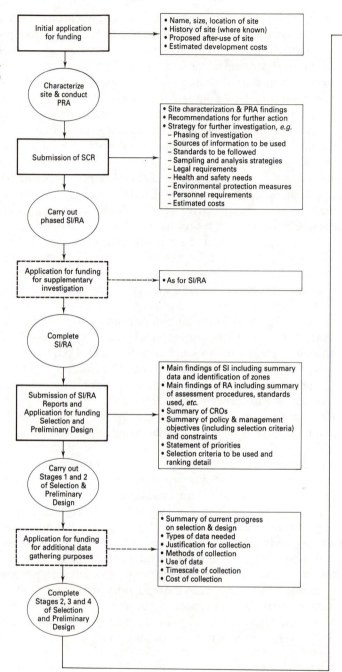

Risk Assessment and Management Strategies

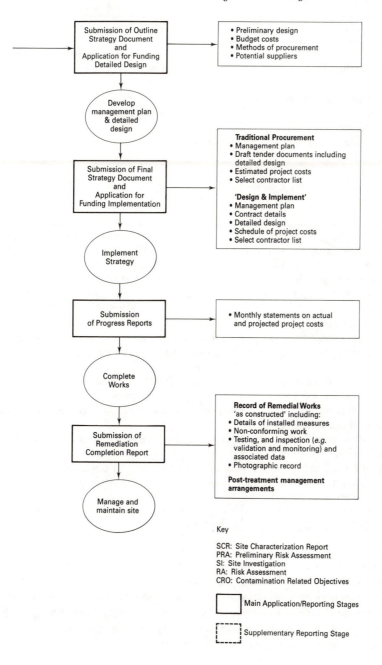

Submission of Outline Strategy Document and Application for Funding Detailed Design
- Preliminary design
- Budget costs
- Methods of procurement
- Potential suppliers

Develop management plan & detailed design

Submission of Final Strategy Document and Application for Funding Implementation

Traditional Procurement
- Management plan
- Draft tender documents including detailed design
- Estimated project costs
- Select contractor list

'Design & Implement'
- Management plan
- Contract details
- Detailed design
- Schedule of project costs
- Select contractor list

Implement Strategy

Submission of Progress Reports
- Monthly statements on actual and projected project costs

Complete Works

Submission of Remediation Completion Report

Record of Remedial Works
'as constructed' including:
- Details of installed measures
- Non-conforming work
- Testing, and inspection (*e.g.* validation and monitoring) and associated data
- Photographic record

Post-treatment management arrangements

Manage and maintain site

Key

SCR: Site Characterization Report
PRA: Preliminary Risk Assessment
SI: Site Investigation
RA: Risk Assessment
CRO: Contamination Related Objectives

☐ Main Application/Reporting Stages

⌐ ¬ Supplementary Reporting Stage
⌊ ⌋

117

Table 1 Examples of final selection criteria*

Long term	Short term
Performance and legal compliance	Acceptable operational requirements
Long-term effectiveness	Minimal short-term health and safety impacts
Reduction in toxicity, mobility and volume	Minimal short-term environmental impacts
Acceptable track record of use	
Acceptability to regulators and local community	

*Welsh Development Agency, 1994.

Table 2 Aspects to be covered in the management plan*

Technical, legal, financial and policy objectives of remediation
Technical requirements of the proposed remedial strategy, including those related to health and environmental protection
Preferred/available procurement and contract options
Resources (personnel, finance, time, materials) to be made available to carry out the works
Roles and responsibilities to be allocated
Lines of communication (both internal and external, *e.g.* with regulators, local community) necessary to achieve safe and effective remediation
Types and levels of contingency to be made
Procedures required to ensure legal compliance, monitor the progress and quality of the works and confirm performance over the long term
Procedures required to validate the work and to enable corrective action to be taken where necessary
Procedure required to ensure effective post-treatment management where appropriate
Systems required for documenting and reporting all objectives, plans, decisions made and actions taken, including changes to the original design concept or agreed methods of working

*Welsh Development Agency, 1994.

to agree a phased programme in situations where different techniques are employed in different areas of the site or early release of a part of the site for development purposes is required. The client needs to recognize, however, that he may not be able to seek redress for the correction of any work subsequently found to be defective. For this reason, the timing of validation and the intensity of sampling needs to be given careful consideration.

4 Site Experience at Queensferry

The benefits of a robust management strategy can perhaps best be illustrated by reference to a specific case study. A former oil and chemical reprocessing site situated on the banks of the River Dee at Queensferry, in North Wales, was first identified for reclamation in the mid 1980s. Initial investigations at the site indicated that the ground was contaminated to a significant degree, but it was not until the site was being vacated in the early 1990s that progress was made in the implementation of a remediation scheme. Although the contaminated nature of the site was known throughout the intervening period, pollution of the adjacent river was not occurring and the contamination appeared to be contained within the site boundaries. Consequently, the companies occupying the site did not suffer any enforcement action to clean up the site or to prevent further contamination—a sad reflection on the limited powers available to regulators and the local authority at that time and on the poor management practices adopted by operators.

The result was that by the time reclamation was contemplated, the severity of the contamination had increased. Pools of oil and stacks of disused drums, with unknown contents, were present on the surface, open tanks were overflowing (principally due to rainfall) and buildings within the site were in a derelict condition; Figure 5 illustrates the condition of the site. In addition, it was known that drums had been buried beneath the surface and chemical residues had been deposited in the ground.

A reappraisal of the site was undertaken by specialist environmental consultants prior to the introduction of procedures from the Agency's Manual, but, nevertheless, it adopted the philosophy of a phased approach. The initial non-invasive investigation was aimed at identifying the full extent of potential hazards through historical review, site reconnaissance and characterization. The subsequent main investigation was, therefore, focused on those elements of the site for which information was lacking, *e.g.* the buried drums, the contents of surface drums and the extent of groundwater pollution.

Evaluation of these data identified potential strategies for treatment, one of which was an innovative chemical dispersion technique. This technique was only available in the UK through one company, which was commissioned to undertake on-site evaluation trials. Unfortunately the trials concluded that the technique was not viable at that stage of its development. The subsequent review of the strategy during 1992 did not have the benefit of a formal assessment procedure and a significant delay (and much debate) was incurred before the re-evaluation was accepted.

The resultant strategy—excavation and disposal of contaminated ground to an off-site tip—was based on the objective of achieving an industrial development end-use for the site. Thus, the CROs were defined by a specialist consultant, who prepared the detailed design and specification for a conventional procurement.

Tenders were sought from a select list of major contractors who were thoroughly vetted prior to their inclusion on the tender list. Commencement of the contract works was coincident with the introduction of the Manual and the consultant adopted monitoring and supervision procedures as required by the document. Control of the works involved the establishment of a 10 m grid pattern across the site with the contamination status of each grid square being established from representative samples. The results of sample analysis were used to define whether the material met the residual target levels. All material exceeding these levels was taken for disposal off site, together with the tanks, drums and their contents. The site was restored using imported, 'clean' fill and capped with

Figure 5 Site at Queensferry prior to remediation

sub-base material to provide a development platform. The oil-contaminated groundwater was subjected to a pump-and-treat system. Figure 6 illustrates the completed site.

The design of the scheme and its subsequent implementation proceeded with the benefit of full consultation with the regulatory bodies (Local Authority Environmental Health Officers and National Rivers Authority). Those bodies were advised of the status of the site at all stages of the project. This liaison was essential in agreeing, in particular, the extent of groundwater treatment undertaken and achieving a favourable validation report. Furthermore, the management and monitoring procedures adopted for the project enabled decisions on the treatment of contamination not identified prior to the contract to be made expeditiously. Thus, the progress of the works was not impeded and contract completion was substantially achieved within the contract period, in spite of the inevitable increase in the quantities of material to be treated. A major factor in this achievement was the quality of the consultants and contractors and their working relationship established on site.

5   Future Challenges

In the light of the 1990 Environmental Protection Act[3] and as the full implications of the 1995 Environment Act[7] emerge, it is worth reviewing the future role of the public sector in dealing with contaminated land. The Environment Act, subject to the form that the proposed guidelines finally take,

[7] *The Environment Act 1995*, HMSO, London.

Figure 6 Site at Queensferry following successful remediation

has moved us forward in two fundamental areas. The first is that a 'Remediation Notice' will be able to be served not only if a site is causing pollution but if there is a potential threat. This still depends on definitions but will increase the scope for taking action which is available to the public sector. Secondly, and perhaps constituting the much greater change, the historic polluter will be required to pay for the pollution or threatened pollution which he has caused. Failure to find such a historic polluter will, of course, still leave potentially onerous burdens with what are very often innocent land owners. Many of the effects of the process of Remediation Notices mirror the impact of the Mines and Quarries (Tips) Act 1969 as it relates to unstable tips and should form a basis for the negotiation of dowries to meet the cost of remediating sites. This could strengthen and make more common the concept of sites having negative value.

A dangerous and overly simple belief which is emerging is that the legislation now in place is such that the polluter will pay for all past ills. This, of course, is not the case. The polluter or owner will pay for the pollution or the threat of pollution for which he has a proven responsibility, but will not be required to prepare the site for regeneration. The work needed to meet these liabilities will often fall far short of effective and comprehensive reclamation of a site to enable its future use, and may even prevent or complicate its reclamation. There is, therefore, going to be an inevitable call upon public sector funds as a grant of last resort, firstly where the historic polluters and land owners are likely to suffer hardship as a result of the cost of remediation and a dangerous situation needs to be reversed, and secondly to carry out comprehensive reclamation of difficult and derelict sites where reclamation of the site is required in the broader public interest and cannot be achieved commercially. As stated previously, however, the Environment Act and the process of a Remediation Notice could well assist the public sector in its reclamation role by strengthening its ability to negotiate a dowry in lieu of short-term and potentially abortive pollution control works.

An area of concern relating to the Environment Act is that the complexity of the problem of identification of the polluter, the pollution which may be caused, and what is an adequate treatment, will rely heavily on the broad-ranging legal interpretation of these factors. Every effort therefore must be made to enable the Environment Act to be implemented in a clear and effective manner and to avoid the trap of spending large sums on legal actions in pursuing scapegoats through the courts rather than on finding solutions.

Whilst the foregoing text sets out the ongoing role of the public sector in the process of the delivery of land reclamation, the public sector also has roles in improving the quality and effectiveness of land reclamation and its implementation and, perhaps even more importantly, must continue to move towards a position where there will not be the need for remediating contaminated land in the future.

There is now developing a clear understanding and acceptance of the adoption of the risk assessment approach in dealing with contaminated land, but there is a dearth of data to enable the risks posed to be defined numerically. There is, therefore, scope for considerable further research into the behaviour of pollutants in terms of their availability, mobility and impact upon specific targets. Many of the answers in relation to health will depend upon epidemiological research and to this end a collection of good and reliable data over a significant time span will be essential, as will be its careful analysis. Continued research in this area is essential to inform the setting of standards relating to tolerable levels of contaminants. It is crucially important that the land reclamation business moves to a set of standards which are internationally accepted and are consistent, credible and sustainable in terms of the financial impact of achieving them, their acceptability and their validity in comparison to naturally occurring background levels.

To improve the quality of what is achieved on contaminated sites, there needs to be further research and development into both destructive techniques and those involving containment of contamination. Both of these technologies are developing steadily but there is obviously further progress to be made. However, to capitalize fully upon these improving and improved techniques it is of over-riding importance that the industry's ability to separate clean material from dirty material is dramatically improved. Currently, identification and separation techniques are the Achilles heel of the industry, in that to ensure removal of dirty material very large volumes of clean material are also frequently lifted and treated. This can result in a waste of resources in handling and treating this mixed material. Furthermore, it means that the best possible treatment for the contaminated material is not affordable because of the exaggerated volumes to be dealt with.

It is extremely important that all developments and innovations are carefully evaluated and proven, because nothing would set the industry back like a major failed experiment. In recognition of the pace of new developments, the Agency will be implementing a regular review of its management strategy to ensure that up-to-date guidance and best practice is adopted in all remediation schemes.

It goes without saying that the fundamental objective of the industry must be to eliminate the need for the remediation of contaminated land. This can hopefully be achieved by improved control, with greater penalties being placed on wilful polluters, and the effective control of the inescapable wastes that continue to be produced. A key step in minimizing the amount of this waste being generated will be for manufactured products to carry the full price of the pollution which their manufacture generates.

Whilst those involved in dealing with contaminated land still face major challenges and must avoid complacency, knowledge, technology and legislation are moving in relative harmony, and considerable comfort can be taken from the progress that has already been made.

Legal Liabilities and Insurance Aspects of Contaminated Land

ANTHONY J. LENNON

1 Introduction

Before consideration can be given to the principles and practices of environmental insurance and the insurance of contaminated land, one must first appreciate the basic concepts of insurance. Insurance is concerned with financial risk transfer. It is an arrangement by which one party (the insurer) promises to pay another party (the insured or policyholder) a sum of money if something happens which causes the insured to suffer a financial loss. The responsibility for paying for such financial losses is then transferred from the policyholder to the insurer. In return for accepting the burden of paying for losses when, and if, they occur, the insurer charges the insured a price—the insurance premium.

The presumption, based on statistical analysis by the insurance company, is that if there are enough people purchasing the class of business, the total pool of premium generated is greater than the cost of the total liabilities encountered during the insured period. Insurance companies can also balance losses on one class of insurance with profits from another, thereby balancing books in any one accounting period. Insurers can also gain income from investing the premium payments and hence earning income on the premiums before any claims may be paid.

Major problems come when the total losses incurred by an insurance company exceed the available income together with any reserves they may have built up over a period of years. This can arise for a number of reasons:

- more of the insured risks occurred than first estimated
- the insurance company did not understand fully the risk they were accepting
- the financial consequences of the risk were far higher than had been estimated by the insurers
- the liability regime understood by the insurer changed from underwriting the risk to when the damage occurred
- the insurer was held liable for costs they had not anticipated, *etc.*

Some of the above reasons have caused the insurance industry to be very circumspect about environmental liabilities and their ability to assess these liabilities effectively in a profitable manner. If the insurers cannot underwrite risks profitably, then they will not have enough funds to pay claims and hence the insured, when the risk occurs, may find that the insurance company is unable to meet their financial liabilities.

Insurance works only because insurers can collect together a large number of similar exposures or risks. Since the markets work with such large numbers of exposures, they can employ statistics to the claims experience and hence work out the probabilities of claims occurring. The larger the number of similar risks that are insured, the more accurate the insurers can be in calculating the likely loss or claim rates and hence developing premiums that provide them with profit and yet are equitable and more truly reflect the probability of the risk occurring and a claim being made against their policies.

The insurance markets, however, apparently do not have the experience or expertise to consider environmental risks in the same way. The question is why, and what is being done to improve the situation for potential insureds?

2 Basis of Insurance Policies

There are two main alternative operating principles for insurance policies: claims made or occurrence wording. An understanding of these two different operating principles is vital to be able to appreciate the insurance industry's concern over environmental liabilities and their apparent inability to assess properly their exposures.

Occurrence Wording. This type of policy format works on the principle that the insured purchases a policy for a fixed period of time (say one year) and the policy will then respond to any claims arising out of actions of the insured that occurred during the period of insurance, no matter how many years later the damage manifests itself.

Claims Made. This type of policy format relies on the loss for which the policy is providing indemnity occurring during the insured period and the claim being lodged against the insurance policy during the insurance period.

Public liability insurance provides indemnity to the insured against their legal liability incurred for property damage and bodily injury to third parties. It normally provides indemnity to the insured for the financial consequences of damage to third parties caused by any activities of the insured giving rise to such damage, which would not be insured under more specific forms of insurance. Such policies in the UK are normally written on an occurrence-worded basis. However, since pollution damage can take many years to manifest itself, any indemnity policy issued on an occurrence wording basis is almost impossible to assess for potential pollution liabilities because of the possibly massive time delay between exposure and the manifestation of damage.

3 History of Environmental Insurance

As with so many things relating to environmental protection, we have to look to the USA for the first real experience the insurance industry had in considering environmental risks. Until the late 1970s the American insurance markets offered public liability insurance on an occurrence-worded format with no pollution restrictions. Frightened by the problems associated with asbestos and the ever tightening legislative system in the USA, the market attempted to restrict cover for pollution-related damages.

Initial attempts were made to place restrictions on claims that were related to gradual pollution. It was soon discovered that for all practical applications it is extremely difficult to differentiate between gradual pollution damage and more instantaneous or sudden and accidental pollution damage. As claims against the partially restricted policies started to flow, it became apparent that the USA courts were not making consistent decisions and interpretation of the pollution exclusions. The consequence of inconsistency, and the protracted litigation cases between the insurers and the insureds, caused the market to respond by placing total pollution exclusions on their public liability policies (known in the USA as Comprehensive General Liability policies, CGL).

The imposition of pollution-restricted CGL policies allowed the development of pollution-specific insurance products that could be underwritten independently of other forms of insurance.

4 USA Experience

The development of environmental policies was not without its problems. By the end of the 1970s there were only two insurers able to offer such policies. By the end of 1983 there were some 40 insurers offering environmental policies. Unfortunately, by the end of 1984 the environmental insurance market had collapsed, leaving only one insurer surviving in the USA market.

The USA market failed so spectacularly because the underwriters did not hold true to their principles and apply sound technical principles when underwriting the risks. In order to underwrite environmental risks effectively the following points need to be considered:

- underwriters should have sound technical knowledge and be able to understand the problems and environmental significance of industrial activities
- the insured should be required to employ high standards of risk management and loss prevention
- sites should be surveyed by competent environmental surveyors prior to insurance being bound
- up-to-date knowledge on the understanding of the properties and significance of substances that may be hazardous to the environment.

Only if such principles are adhered to can environmental risks be underwritten successfully.

5 UK Insurance Response to Environmental Liabilities

UK insurers' initial response to the problems of environmental liabilities and their effects on them was largely to ignore them and pretend that they did not exist. This lack of realization has now started to change and in 1991 the general UK insurance market clarified the cover they were providing by restricting cover to sudden and accidental incidents leading to pollution-related damage by employing exclusion wording similar to:

'This policy excludes all liability in respect of pollution or contamination other than caused by a sudden identifiable unintended and unexpected incident which takes place in its entirety at a specific time and place during the period of insurance'.

Pollution Exclusion

The 'gradual pollution exclusion' has now become almost universal on all public liability policies, but what does it mean? Unfortunately the UK insurance industry has learnt nothing from the experience in the USA during the 1980s and has fallen into the trap of thinking that it can restrict effectively its exposure to environmental liabilities by limiting cover to sudden and accidental incidents. It has been learnt in the USA that the concept of sudden and accidental pollution damage is very difficult to interpret effectively and consistently. It is not clear from the words used what cover is being provided. A simple example will be sufficient to illustrate the point. An explosion attributable to methane gas from a landfill site is quite obviously a sudden and accidental incident. Unfortunately, however, in deciding a claim the insurance markets will not look for what may appear as the obvious cause of any damage, but will seek to find the root or proximate cause leading to the damage. In this example the proximate cause could be attributed to any of the following:

- gradual build up of methane in the space prior to ignition
- gradual decomposition of the waste in the landfill
- gradual seepage of methane from the landfill
- gradual deposit of waste in the landfill

As can be seen, most of the above could be considered as gradual and hence would probably be excluded from cover from a UK public liability insurance policy. The issue in this case is not actually which of the causes would be considered as the proximate cause, it is rather that there are a number of potential causes, most of which may be considered as gradual and that the exclusion wording is ambiguous and allows interpretation. The other point to consider in the above exclusion wording is that it relates to an incident leading to damage and not the damage itself.

It has been reported recently that it is recognized by many that the current pollution exclusion is open to debate and that certain sectors of UK industry have developed a fund to test the exclusion wording in court.[1] Although not documented, there are significant indications in the UK insurance market that it will not be long before UK public liability insurance policies will have total pollution exclusions in the same way as CGL policies do in the USA.

6 Land Owners and Environmental Liabilities

Owners of land face potential liabilities with respect to environmental damage, from both the past and present land uses. It is often thought, incorrectly, that the

---

[1] *Composite Insurance, the Environmental Pollution Threat*, James Capel, UK Equity Research, London, 1993.

primary liabilities facing land owners are those associated with past land use. Past land use does give rise to substantial potential liabilities, although after the recent House of Lords decision concerning the case of Cambridge Water *vs.* Eastern Counties Leather, the civil liability land owners may face for past contamination may not be as great as was initially thought.

Although potential liabilities for damages caused by contaminated land and past industrial activities have been addressed by the House of Lords, the costs of cleaning contaminated land up, and the acceptability of that land once cleansed to potential land purchasers, is still a very real issue. What should be of equal concern to land owners, particularly those that lease land to others (for example, industrial estates), is the potential for the site to be contaminated by current land uses. Linked with these are the liabilities that may fall on land owners for contamination to land and damages to third parties as a result of today's activities rather than historical uses. Each issue is dealt with separately below.

Contaminated Land

The redevelopment of contaminated land has always been relatively big business. Over the last couple of years it has been facing a number of significant problems:

- possible implementation of the registers of potentially contaminated land as required by section 143 of the Environmental Protection Act 1990
- more recently, the considerable discussion and speculation concerning the implementation of the contaminated land liability regime contained within the Environment Act 1995
- the continual delay that has dogged the production of the statutory guidance to local authorities that needs to be in place before the Environment Act liability regime can be implemented
- falling land values and interest in land development
- inability of valuers to value effectively land which may be/is contaminated, *etc.*

These problems, linked with the ever increasing public interest in the environment and its protection, have meant that contaminated land development has not been easy. Lending institutions and banks have also started a harder line towards lending money for such developments. Some developers have been faced with the situation that, although they have found sites that can be decontaminated and developed profitably, banks have refused to lend any money because the site is contaminated.

The current planning philosophy is to promote development on so-called brownfield sites—those that have had previous uses—and there needs to be a distinction between remediation of contamination as a result of land development and that as a result of regulator intervention. The recently introduced contaminated land liability regime introduced in the Environment Act 1995 is designed to consider land that does or is likely to cause harm during its current use. The presumption is that the vast majority of land decontamination will be undertaken outside the statutory regime and will be done as a normal part of land development.

It is now usual to have land investigated by a competent environmental consultant who has experience in the necessary areas of work. A very worrying and misguided notion appears to be spreading through land developers: they appear to think that if any contamination is found after decontamination, or if any problems develop with the site, they will be able to sue the consultant.

The Size of the Problem

It is surprising that within the UK there is no reliable information on the number of contaminated land sites there may be, or even how many of them may be causing harm to the environment. In a recent edition of *Environmental Data Services*[2] it is reported that a so-far unpublished National Rivers Authority (NRA, now part of the Environment Agency) document identifies that, within the Severn Trent NRA region alone, there are 186 contaminated land sites known to be, or suspected of, causing pollution to surface or groundwaters.

The problem of identifying how many contaminated land sites there may be in the UK is further exacerbated by the simple fact that we do not have any clear definition of what constitutes a contaminated land site. In the UK there are no land quality standards against which a site can be objectively tested. The best we have are the guideline figures produced by the Interdepartmental Committee on the Redevelopment of Contaminated Land (ICRCL). The 'standards' give guidance on what quality of land may be required to support a particular end use (*e.g.* housing), but they are not absolute standards that can be applied to a site to declare authoritatively the site clean or decontaminated.

There are proposals and presumptions that the UK Department of the Environment will publish some form of standards or guideline figures to be used to assess contaminated land. Such figures are likely, however, to be little more than indicative numbers that developers can use to consider the levels of certain substances and their effects on humans. The general presumption under the new contaminated liability regime will still be that a developer will undertake a risk assessment of the contamination that may be present on their land and what amount of decontamination will be required in order to reduce the risk of the contamination giving rise to problems to a minimum. Although such a system passes the responsibility to identify the level of reclamation to those that are likely to implement the decontamination, it could be argued that such a system will promote further concern because land owners will not be able to look at lists of chemicals and the defined or accepted safe level for them in soils.

Pollution Liabilities for Contaminated Land Sites

It must be appreciated that contaminated land sites can exhibit pollution liabilities of two basic forms. These can be liabilities arising from pollution to people and property off site—so called third party liabilities—and also liabilities relating to pollution of the site itself and the consequential need to clean up the site. The second liability may, and in some cases is likely to, lead to damage to people or property off site.

[2] *The ENDS Report*, 1994, No. 230, 23.

As discussed above, damage caused to a third party's interests from contamination migrating off a contaminated site is unlikely to be covered by a current public liability policy. Any migration would almost certainly be judged by an insurer to be gradual in nature. In this situation the only hope for the owner of the land is either: (i) to have a specific pollution liability policy for the site; or (ii) to hope that the migration can be traced back to a time when the current owners had a public liability policy without a pollution exclusion and to hope that the insurance policy was written on an occurrence worded basis; as time progresses, however, this is likely to become increasingly difficult because of the difficulty in being able to link the pollution damage with an incident that occurred away in the mists of history.

Costs to clean up the insured's own sites are not normally available from the traditional insurance market. It is not a claim that would be considered under a current public liability policy and if the contamination was known about, then the site could not hope to obtain site specific environmental insurance to cover the required clean-up costs. The only situation where there may be a glimmer of hope is in situations where those facing liability for clean-up were tenants on the land and not the land owners themselves. It was, and still is, relatively common practice for tenants to have their public liability policies extended to provide indemnity to them for any damage they caused to the freeholder's land or property. If such a policy extension were to be available on an old occurrence-worded policy held by a tenant, the property damage part of the cover may provide them with some opportunity to gain an amount of insurance indemnity for any requirement placed upon them to clean up their landlord's land.

Most land owners and tenants hold property policies which provide indemnity to the insured for the costs of rebuilding the premises as a result of fire or other named perils. These policies also provide indemnity to the insured for debris removal on site arising out of a specified named peril, *e.g.* a fire. These policies, although providing a degree of own site clean-up cover, would only respond to contamination and debris removal arising out of the named peril and would, therefore, not respond to the general removal of contamination on the insured's premises.

Current Manufacturing Activities

'Manufacturing industry today is over-regulated and hence there is no potential for manufacturing industry to give rise to either land contamination or damages to third parties'. Is this a valid comment? In my view, it is most certainly not for a number of very important reasons:

- not all potentially polluting processes are regulated
- regulation is still largely based on media interests, *e.g.* water pollution, air pollution. Because of this, it is quite conceivable that over-regulation for one environmental medium may cause a transfer of pollution potential to another environmental medium. To some extent this has been addressed under the principle of Integrated Pollution Control, but this form of regulation only encompasses a very small proportion of manufacturing activities in the UK

- regulation tends to concentrate on normal operations in a manufacturing activity and does not always address potential aberrations in operational practices
- the sanction in the event of a failure to comply with legislation is usually no more than a fine and very rarely does it encompass compensation or other means of providing restitution to injured parties or the environment

For all the above reasons and more, I believe that regulation in the UK and throughout the rest of the world does not guarantee that a regulated manufacturing process will never cause damage to the environment. In the 1994 Annual Report of the NRA, covering their activities during 1993, it was recorded that there were over 30 000 reported water pollution incidents in England and Wales throughout 1993. Although this number is likely to include a large proportion of relatively minor incidents, every one of them could have the potential of causing significant damage to the aquatic system and hence damage to other people's interests and property. The NRA's final report in 1995 recorded that although the number of serious pollution incidents had fallen, the number of reported water pollution incidents in 1994 had increased to over 35 000.

The number of complaints received by Her Majesty's Inspectorate of Pollution (HMIP, now part of the Environment Agency) is also on the increase, again demonstrating that regulation in itself will not always result in satisfactory operational characteristics of a plant.

One simple example of where current manufacturing activities have given rise to contamination will suffice to demonstrate that today's activities can and do cause pollution and contamination. A small industrial estate in North London was owned by a company that let the site out as an industrial estate. One of the units was let to a metal electroplating company. The company went bankrupt and stripped all the valuable materials and equipment from the site one night and left the land owner with the problem of disposing of all waste materials that were left, together with a significant pollution legacy. The brick walls of the building were so badly corroded with plating chemicals that it would have been possible to push a hand through them without encountering any significant resistance from the brickwork. The floor and underlying ground were also grossly contaminated by the plating processes that had only ceased operation a matter of days before. The land owner was faced with removing and cleaning his site since he was unable to obtain any restitution from the occupiers of his industrial unit. There was also the problem of whether or not the contamination had affected the integrity of any surrounding units that he owned.

It must be appreciated, recognized and always remembered that manufacturing activities we assume are benign often are anything but that, and that they may have as much potential for causing damage to the environment as the predecessor processes operating in Victorian or earlier times.

7 Environmental Insurance

As has been mentioned before, Public Liability policies, in the main, have some form of pollution exclusions attached to them. Property Damage policies, which provide indemnity in the event of damage being caused by named perils, only provide indemnity for removing contamination from that portion of a site actually affected by the named peril. For example, a large site which has a fire in one building yet causes contamination over the whole site will not have indemnity to pay for the clean-up associated with the contamination from the fire from any land other than that immediately surrounding the fire. It is very unlikely that a large site contaminated by a fire would have any adequate insurance indemnity available to decontaminate it from pollution caused by a fire or other perils. Because existing policies have significant restrictions placed upon them in terms of the degree of cover provided for pollution-related liabilities and clean-up costs, two particular policies have been developed to indemnify insureds for these liabilities; these are pollution legal liability insurance and own site clean-up insurance.

Pollution Legal Liability Insurance

This provides indemnity to the insured for damage caused to third parties as a result of pollution conditions emanating from the insured's premises. It has been used extensively in the USA and is now available in the UK. Some companies refer to this type of policy as Environmental Impairment Liability insurance. This title is a slight misnomer because the policy does not indemnify the insured for environmental damage *per se*, but for bodily injury and property damage resulting from pollution caused by the insured.

Own Site Clean-up Insurance

Within the UK, as in the USA, the pollution regulators have a series of powers which are either available to them now or are currently in the existing legislation but are awaiting implementation. All of the following UK regulators have powers that could require a site owner to clean up their site:

- National Rivers Authority
- Her Majesty's Inspectorate of Pollution   } now part of the
- Waste Regulation Authority   Environment Agency
- Health & Safety Executive
- Local Authority

There is an insurance policy available now that will indemnify the insured for the costs incurred, up to the limit of indemnity, in the event of any pollution regulator using their powers to institute a clean-up of their site. This is a particularly important policy since it is quite common for a third-party pollution claim to lead to a clean-up requirement on the site giving rise to the problem, and the current public liability policies will not provide any indemnity for such expenses.

A recent development in site-specific insurance is the packaging of the two types of cover into a single policy. This has the advantage that the insured has only one policy to concern themselves over and that all the insured risks are adequately covered under one single and simple form, with less chance of an uninsured risk developing.

Benefits of Environmental Insurance

There are numerous reasons why such policies should be considered, not just for reclaimed contaminated land sites, but also for sites where contamination is not suspected. Some obvious reasons are:

- unknown past and future contamination/pollution damage is covered
- in their investigations, consultants may miss contamination
- reclamation strategy may not be robust enough to deal with all contamination found
- there may be insufficient information on past land use
- perceptions against contaminated land may harden
- contamination may be present even though none is found at investigation
- ongoing land uses may cause additional contamination after reclamation
- reclaimed contaminated sites that have the benefit of site-specific environmental insurance may be more attractive to investors and end users
- banks and other lenders can be added onto the policies to note their interest in the property
- land owners can also be added to the environmental policies to ensure that should their tenants at any time become bankrupt or if they ran away from their liabilities, there is at least some insurance indemnity available to provide the land owner with protection against the financial consequences of these environmental liabilities
- policies can be written and provided on a multi-year basis

General Features of Environmental Insurance Policies

The main features associated with these policies are:

- written on a claims made basis
- site specific
- subject to survey
- provides indemnity whether caused by sudden and accidental events or gradual events
- sites individually underwritten and location covered are specifically identified on the policies

These policies have been developed in the USA and in their early stages of development reflected the needs of American industry and the legislative framework within which it is required to operate. They are equally applicable to a wide range of applications in the UK. As can be seen above, the policies do not differentiate between sudden and accidental or gradual causes of pollution damage. The policies do not have a specific retroactive date, so they can also be used to provide indemnity to the insured for claims arising out of unknown past contamination.

The majority of industry in the UK has a degree of pollution liability risk, the exact nature and extent obviously being dependent on the operations undertaken on the site. Manufacturing industry, in particular, should be aware of the pollution exclusions already on their public liability insurance policies and should analyse the potential risks they have for either first- or third-party pollution liabilities. At the final analysis, captains of industry should not be asking themselves whether they can afford to purchase these new policies, rather they should be considering whether they can really afford to continue operating their factories without the protection that these new policies can now provide.

8 Environmental Consultants

A survey of a contaminated land site, together with associated sampling and analysis, may cost hundreds of thousands of pounds. The remediation of the site as a result of the consultant's sampling, analysis and advice may cost many millions of pounds. The value of the land, once it has been developed, may be many tens or indeed hundreds of millions of pounds. If the consultancy were negligent in the way in which it took the samples, caused the analysis to be undertaken, in its assessment of the results of the analysis, or in the way in which it proposed the site should be cleaned up, and as a result the site were to be faced with a further clean-up bill of millions of pounds, therefore making the site worthless in the short term, who would have to pay?

It is an accepted fact that consultants owe a duty of care to their clients. A consultant's clients expect a quality of work above that which would normally be supplied by an average person with average knowledge. Consultancies are expert in their chosen areas of work and therefore the client can expect a quality of work commensurate with that professional standing and position. Unlike the USA, however, this duty of care is only shown to the consultant's primary client and the UK consultant does not normally need to consider any further users of their work. In order to limit their liability, American consultancies tend to place limitations of liability either within standard contract conditions or within their reports, making it quite clear that their reports should be utilized only by their primary client and for the original purpose of the consultant's brief.

In order for a client to seek damages from a consultant successfully, they must first prove that the damages they have faced were as a result of the consultant's negligence. This must be proven on the balance of probabilities. It is quite conceivable, however, that contamination may well have been missed not because the consultant was negligent, but because of unsuspected contamination that could not possibly have been suspected or found by the consultant.

Collateral Warranties

In order for consultancies to allow their reports to be utilized by people other than their primary clients, it is not uncommon in the UK for consultants to enter into what is known as a 'collateral warranty'. Although such a warranty may have other uses, one of its features is to allow the consultant's report to be utilized by an organization other than the primary client. For example, a developer may wish to purchase a plot of land and utilize a consultant to identify any potential environmental issues. The consultant produces a report and on the basis of the report the developer purchases the land. Having developed the site, the developer then wishes to sell the land off to a third party, but the third party does not wish to have a survey of the site undertaken for his own benefit, but wishes to rely upon the developer's initial survey. In such circumstances the developer, the third party purchaser and the consultant enter into a collateral warranty, thereby allowing the third party purchaser rights to the consultant's report.

Insurers have to be happy that the collateral warranty is worded in such a way as to provide them with an adequate degree of protection in the event of the third party taking any claims for negligence against the consultant. If entered into blindly, collateral warranties can extend significantly the exposure of insurers in the UK to claims that would normally be outside the scope of common law or contractual responsibilities. However, does a collateral warranty have any real value? A collateral warranty is only as good as the insurance that is in effect at the time of any claim being made, rather than being dependent upon the consultant's insurance that is in force when the collateral warranty is drawn up. With the potential for professional indemnity insurance in the future being limited such that it will not cover any claims relating to pollution or contamination, any collateral warranty entered into in the past may have its future value seriously undermined.

Limitation of Liability

A consultant can be subject to two forms of liability: liability under contract law and for claims relating to negligence and brought under tort.

In the US it is very common for consultants to attempt to limit their liability through conditions in their contracts. The liability is often limited to the value of the contract undertaken or some other amount specified within the contract conditions. There has been a degree of debate about whether or not a condition limiting liability in a contract would also limit their liability of a claim were taken for negligence under tort. It is now becoming more common for American environmental consultancies to place within their contract a condition limiting not only liability under contract, but extending that liability to any other claims for damages howsoever brought against the consultancy.

The practice of limiting liability in contracts is not common in the UK, although it is becoming more and more prevalent and is something that is not uncommon in consultants' contracts where the consultancy has any experience of work within the USA. The contract condition limiting liability has been tested at various times within the United States[3] and has been upheld particularly in situations where the contract value is only a very small proportion of the claim being brought against the consultant. This, for example, could be in a situation where a £25 000 consultant's report is used as the basis for a multi-million pound decision to purchase any individual plot of land. Whether or not such conditions would be upheld in a UK court is open to debate and conjecture. It is believed that to date there have been no such cases brought in the UK.

The market for professional indemnity (PI) insurance is hardening against

---

[3] For example, in 1992, California Court of Appeals in MARKBOROUGH CALIFORNIA, INC. V. SUPERIOR COURT, 227 Cal. App 3d 705, 277 Cal Rptr. 919 (4th Dist. 1991).

environmental consultancies. Why? Probably because traditional PI insurers do not really understand the role of the environmental consultants and are scared by the magnitude of their potential liabilities.

More and more insurers are wishing to place some form of pollution liability exclusion on PI cover. The type of wording used varies from insurer to insurer and, in some cases, some insurers have attempted to exclude any claims attributable to or arising out of gradual pollution incidents. The form of words used in a Lloyd's policy recently seen serve as an example: 'Underwriter shall not be liable to indemnify the Assured against any claim based upon, arising out of or relating directly or indirectly from or in consequence of or in any way involving, seepage, pollution or contamination of any kind'. What this form of words means when applied to a PI policy is very unclear, but on the face of it the underwriters are removing all indemnity for any environmental related claim that is associated with pollution. If this is correct then the PI cover is of very little use.

Some insurers have used the type of wording found associated with public liability policies to restrict indemnity for sudden and accidental pollution only, to qualify indemnity for PI policies. Does the concept of sudden and accidental incident have any meaning when applied to a consultant's PI policy?

PI insurance with any form of pollution exclusion for environmental consultants is not worth the paper it is printed on. Any consultancy that is involved in environmental consultancy work must have the security of knowing that they do not have such an exclusion on their policy. It must be questioned whether a PI policy with any sort of pollution exclusion will provide a sufficient degree of indemnity should the worst happen and a claim is made against the insured and then hence against their PI insurance.

Environmental consultants tend not to have a very large capital base. Their greatest resource is their people. It is not uncommon for consultancies to rent out office space. In the event of a serious claim against them they can easily become insolvent. The provision of properly worded professional indemnity insurance, offering cover for all negligence and associated claims, is the only way a consultancy can operate effectively in today's market place.

At the moment, with very hard work on behalf of the broker, it is still possible in the general market to obtain PI cover for large well-established consultancies. Some consultancies have been used to obtaining PI cover on an each and every claim basis, *i.e.* with no upper limit on the aggregate level of indemnity during the policy period. This is a recipe for disaster and insurers are likely to follow the USA experience and provide PI policies with quite clear aggregate limits.

With environmental consultants facing the potential of an ever growing liability potential, this is the only way PI cover can be sensibly and profitably written. In the same way as environmental liabilities can realistically only be underwritten by specialists that understand the problems and environmental significance of industrial activities, so too should PI for consultants be underwritten. The UK insurance market is now seeing a new type of underwriting for PI which is undertaken by technically qualified and experienced underwriters who understand the work undertaken by consultants and the true potential liabilities they face, particularly when they undertake consultancy for contaminated land sites.

Those that use consultancies will have to accept that the opportunity to require them to have such flexible and comprehensive PI policies is now coming to an end. It is more important to ensure that a consultant has a PI policy covering potential pollution-related claims than one that has very high limits of indemnity, because pollution exclusions may in reality only provide indemnity for a very small proportion of the consultant's activities.

9   Laboratories

Quite often the work of an environmental consultant is dependent on good sampling together with professional and efficient analytical work. Consultants could be considered negligent if they themselves do not use laboratory services that provide a highly professional service. It is not uncommon, however, for laboratories to have a very limited, if indeed any, professional indemnity insurance. The relationships between consultant and laboratory are very important and the limitations of the analytical techniques need to be taken into account when the samples are taken. It is sometimes very difficult to distinguish between the failure of a consultant and that of a laboratory.

There is a series of accreditation schemes that laboratories can take part in. Most accreditation systems, however, only relate to certain analytical techniques and it is unusual for them to cover all of the analytical techniques that are offered by individual consultancies or laboratories. It is, therefore, difficult and indeed wrong to assume that an accredited laboratory will never make a mistake or, for example, that if they are accredited for cadmium analysis they can undertake dioxin analysis to the same degree of proficiency.

For those laboratories that hold PI insurance, it is not uncommon for them to have their policies specifically endorsed to exclude all claims relating to pollution. In the same way as for environmental consultants, laboratories which are involved in analysing need indemnification insurance in order to cover all potential claims, particularly those relating to pollution. It is therefore inappropriate for laboratories that engage in analysing environmentally based samples to have professional indemnity insurance that excludes claims relating to environmental pollution or contamination.

10   Contractors

Consultants are engaging themselves in contracting services. This is particularly apparent where a consultancy undertakes a project management role on sites that require decontamination. Should there be any damage caused to the site or surrounding neighbours as a result of the consultant's contracting services, then they would be relying upon their public liability policy to provide them with the necessary level of indemnification. As is quite common, of course, their public liability policy will have a gradual, if not a total, pollution exclusion. If the damage is related to any pollution conditions arising out of any of the contracting services provided by the consultant, then it is possible that they will not have the necessary degree of insurance protection to cover the cost of the claims.

It is not uncommon for contractors to be held liable for any damage as a result of their activities. Where contractors are dealing with hazardous substances, their clients almost certainly require them to assume responsibility for damage caused whilst handling or dealing with the hazardous materials. How can contractors accept such liability with a public liability policy that contains a pollution exclusion?

There is now the opportunity for contractors to obtain specific pollution indemnity insurance to supplement their public liability policies. In essence, these policies provide indemnity to contractors should their activities, as a result of pollution conditions, give rise to property damage or bodily injury to third parties. The main features of these policies are as follows:

- sudden and gradual coverage for third party liability, including defence costs
- claims made policy; retroactive date provision
- three-year extended reporting period option
- completed operations coverage available
- insured's clients may be covered as additional insureds under the policy
- covered operations must be specified on the policy

Where consultants are involved in providing contracting services as well as consulting services, they must ensure that they have an adequate degree of insurance to cover both aspects of their work. Although not common, it is possible to obtain tailor-made policies that encompass the indemnification provisions of both a professional indemnity policy and a public liability policy, providing the necessary degree of cover for pollution-related claims.

11 Conclusion

Public liability policies are already having their scope reduced so that there is no intended cover for gradual pollution damage and there is a very real possibility that their scope will be further reduced in the near future to remove cover for all forms of pollution-related damage. UK industry should be aware of this restriction and should consider whether or not their existing portfolio of insurance provides adequate protection.

Developers of contaminated land should understand the relationship they have with environmental consultants and should not consider the contract they have with the consultants as a surrogate for clean-up insurance.

There are now specific insurance policies available for industry and owners of land to indemnify them for a wide range of their potential environmental liabilities. These policies can be readily applied to development sites once they have been cleaned up to the underwriter's satisfaction. For decontaminated contaminated sites, location-specific environmental insurance can remove some of the uncertainties from the site and promote further investment and development.

Subject index